Welcome to Subirdia

Yale
UNIVERSITY
PRESS
New Haven and London

WELCOME TO

Subirdia

Sharing Our Neighborhoods with Wrens, Robins,
Woodpeckers, and Other Wildlife

John M. Marzluff

Illustrations by Jack DeLap

Published with assistance from the foundation established in memory of
Calvin Chapin of the Class of 1788, Yale College.

Yale University Press books may be purchased in quantity for educational, business,
or promotional use. For information, please e-mail sales.press@yale.edu (U.S. office)
or sales@yaleup.co.uk (U.K. office).

Designed by Sonia Shannon and set in Bulmer type by Westchester Book Group.
Printed in the United States of America.

Library of Congress Cataloging-in-Publication Data

Marzluff, John M.
Welcome to subirdia : sharing our neighborhoods with wrens, robins, woodpeckers,
and other wildlife / John M. Marzluff ; illustrations by Jack DeLap.
pages cm
Includes bibliographical references and index.
ISBN 978-0-300-19707-5 (cloth : alk. paper)   1. Bird watching—Washington
(State)—Seattle.   2. Birds—Washington (State)—Seattle—Identification.
3. Birds—Habitat.   4. Bird watchers—Anecdotes.   5. Urban animals.   I. Title.
QL677.5.M38 2014
598.072′3479777′2—dc23

2014012257

A catalogue record for this book is available from the British Library.

This paper meets the requirements of ANSI/NISO Z39.48-1992 (Permanence of Paper).

10 9 8 7 6 5 4 3 2 1

For the students who sacrificed to learn, the colleagues who generously explained, and the good neighbors who share their land with wildlife

Contents

Preface

IN MY YARD, REDUCING TURF and planting native shrubs bring the song of Swainson's thrushes each spring, while also discouraging the presence of nonnative starlings. Three pairs of spotted towhees nest under the ferns and shepherd forth stubby-tailed, earth-toned fledglings each summer to sample what our bird feeders provide. Thinning a crowded forest canopy spurs tree growth, creates standing dead trees, and supplies downed wood to the forest floor. Western tanagers, pileated woodpeckers, and delicious chanterelle mushrooms approve of these actions. Native Douglas squirrels clamber in the canopy for fir cones, which they clip and drop with a bang onto the deck. In contrast, the junco that weaves a nest of dog hair into the soil of a potted plant is perturbed by our every move.

The ability of birds to thrive alongside humans is the latest testament to their adaptability. Birds have flourished on Earth for nearly 170 million years. As skilled fliers they dispersed across the globe and carved niches from the ground, the canopies of trees, and the sky. The radiation of birds, which today enriches Earth with more than ten thousand species, was not without setbacks. Sixty-five million years ago an asteroid slammed into the planet, extinguishing the dynasty of toothed birds known as the Enantiornithes and Hesperornithes. They had been as common as herons are today, flying among Mesozoic trees and fishing within primordial seas. Despite this lost legacy, the roots of today's modern bird diversity persisted even as their dinosaurian ancestors perished. The agility of those in the surviving lineages allowed birds to capitalize on opportunities afforded by the advances and retreats of continental ice sheets, the extinction of large reptiles and mammals, and now the domination of Earth by urban humans. But with each opportunity also come challenges. Today, our actions threaten many birds.

Ten percent of modern birds are at risk of extinction because of our enterprise. Human conversion of native forests, shrublands, prairies, and coastal marshes into croplands and settlements is the primary reason why so many birds, as well as other forms of life, are threatened. We directly endanger birds by exposing them to new diseases, such as West Nile virus, and overharvesting those that we find tasty, like the passenger pigeon. Climate change in response to our burning of fossil fuels affects plants and the birds that depend on them. Even subtle actions challenge birds. Creating picnic areas, hiking trails, and ski areas in wild country promotes some aggressive species, such as crows and jays, which in turn prey on other birds. On islands, where space is limited and unique animals have evolved in isolation from their mainland relatives, these threats are extreme. Visitors to Hawaii, New Zealand, or islands in the Caribbean may see many beautiful birds, but few are native.

My research and that of other urban ecologists suggest that, despite the great loss of biodiversity caused by our actions, we also have a lot to celebrate. I've spent most of my spring and summer mornings counting birds in national parks, industrial parks, and suburbs. It is not surprising that the most heavily paved portions of the city hold few birds, but it is not the case that the least disturbed places on Earth always hold the most birds. Wild reserves provide shelter for unique birds not found in the city, and they are absolutely essential. But the greatest variety of birds is often found in the suburbs. Some of this variety is otherworldly. The escaped, nonnative red-masked parakeets that colonized San Francisco's Telegraph Hill, for example, are clearly out of place, but dazzling.¹ Some people may question their suitability as replacements for birds lost when coastal forests were developed. But to me, the raucous squawk of a parrot is better than never again hearing the song of an extinct species.

Occasionally, even uncommon birds find city life to their liking. The rare wheatear, a dapper gray bird with a black bandito mask, sometimes forgoes its typical alpine meadow and tundra haunts for the abandoned rail yards and airports of Berlin. These urban pioneers are extremely successful at nesting where trains and planes no longer tarry, but they are an exception. Other rare species shun the city. Prairie-chickens and sage-grouse need unpaved grasslands and shrublands. They have no use for the trees we plant. Spotted owls and Hawaiian crows require large tracts of wild forest. The cities we create and the nonnative species we attract are to these forest specialists no different than the asteroid was to the toothed birds of the Mesozoic.

A growing list of books review and advance the emerging field of urban ecology. *Welcome to Subirdia* complements these efforts by distilling from urban ecological research the important insights, illustrating them with research on birds and making this wealth of knowledge available to bird enthusiasts as well as students and scholars of ornithology, ecology, urban planning, and landscape architecture. The notes and references allow you to engage the literature and explore the abundance of information on urban birds that exists on the Internet.

As I write I am distracted by a sizeable flock of juncos that eats seeds from my feeder, undeterred by a steady, cold rain. Throughout North America the junco, a slate-colored, sparrow-sized bird with a flashy white-sided tail, rules the urban jungle. My aim is to help us understand why. As we will learn, the secret to the junco's success is mostly luck. By chance, the way we mix vegetation, spread grass, dig ditches, and provision bird feeders perfectly suits the junco's requirements. We give it what it needs, and in response the junco population booms. And here is the key to its survival: its large populations enable natural selection to evolve the junco's behavior and physique to match the challenges we humans pose. There are also unfortunate birds, however, and as we celebrate the winners, we will also learn from the losers and consider how we might conserve the birds that struggle to coexist with us.

Ten principles, what I like to think of as Nature's Ten Commandments, can help us appreciate and sustain the birds and other wild animals that live among us (see Chapters 9 and 10). Embracing these commandments can guide your development of a land ethic that holds at its core an appreciation for the community, not just the commodity, of your property. Aldo Leopold encouraged us in such practices more than sixty years ago, and with today's growing urban population, his call is all the more relevant. With this ethic we can reap services and values from nature such as pollination of important crop plants, beauty, and health. Without it, we will continue to distance ourselves from the natural world and the benefits it provides our enterprise, culture, and well-being.

Acknowledgments

THE RESEARCH AND INSIGHTS you will encounter in this book would not have been possible without the dedication of my postdocs, graduate students, and their assistants and the scholarship and collegiality of my professorial partners in urban ecology from Seattle and Berlin. Tina (Rohila) Blewett, Barbara Clucas, Heather Cornell, Jack DeLap, Roarke Donnelly, Laura Farwell, Jeff Hepinstall, Cara Ianni, Sonja Kübler, Peter Meffert, Dave Oleyar, Lin Robinson, Stan Rullman, Ben Shryock, Jorge Tomasevic, Thomas Unfried, Kara Whittaker, and John Withey awoke early and worked late to collect and analyze the data that underpin my understanding of our urban ecosystems. Michael Abs, Marina Alberti, Gordon Bradley, Wilfried Endlicher, Kern Ewing, Ingo Kowarik, Maciej Luniak, Pete Marra, Marc Miller, Maresi Nerad, Amanda Rodewald, Clare Ryan, Eric Shulenberger, Ute Simon, Gerd Wessolek, and Craig ZumBrunnen shared their expertise, enthusiasm, and fellowship as they broadened my understanding of the urban world. Their collaboration enabled successful proposals that funded our joint research from the U.S. National Science Foundation (DEB-9875041, IGERT-0114351, BCS 0120024, and BCS 0508002), the German Research Foundation (RTG Graduiertenkolleg 780), and the University of Washington (Rachel Woods Endowed Graduate Program).

I am thankful to Sam Fox, Trena Keating, Glenn Rimbey, Steward Pickett, Colleen Marzluff, Jessie Dolch, and Margaret Otzel for their careful reads and ruthless pens. Each read the entire manuscript and provided the balance of constructive criticism and encouragement that I needed. Brian Kertson, Pam Yeh, Steve Humphrey, Jeff Norris, David Peterson, Al Sanford, Robbert Snep, Kaeli Swift, Karl Wirsing, Scott Loss, Travis Longcore, Steve West, Amy Yahnke, Rob Blair, Jip Kooijmans, Rob Faucett, and Sarah Sloane clarified important details. Catherine Connors and Zoe Marzluff researched the Latin roots of new scientific names. Mark Mead and Lisa Ciecko provided details on Seattle's trees. Jack DeLap brought our observations to life with his drawings. Trena Keating and the staff at Union Literary honed my thoughts and steered me through the complex world of publishing. Jean Thomson Black and her staff at Yale University Press are good friends and superb editors who encouraged and shepherded my ideas through to publication. The School of Environmental and Forest Science's Ridgeway Endowed Professorship provided me the funds to finalize this manuscript.

My wife, Colleen, and daughters, Zoe and Danika, never fail to amaze me with their patience and support. I appreciate and treasure every moment we share as a family. I am forever grateful to my father, who encouraged and supported my interest in wildlife, even sacrificing his career so that we could together enjoy a suburban life connected to nature. I hope that my efforts equally encourage my daughters.

Welcome to Subirdia

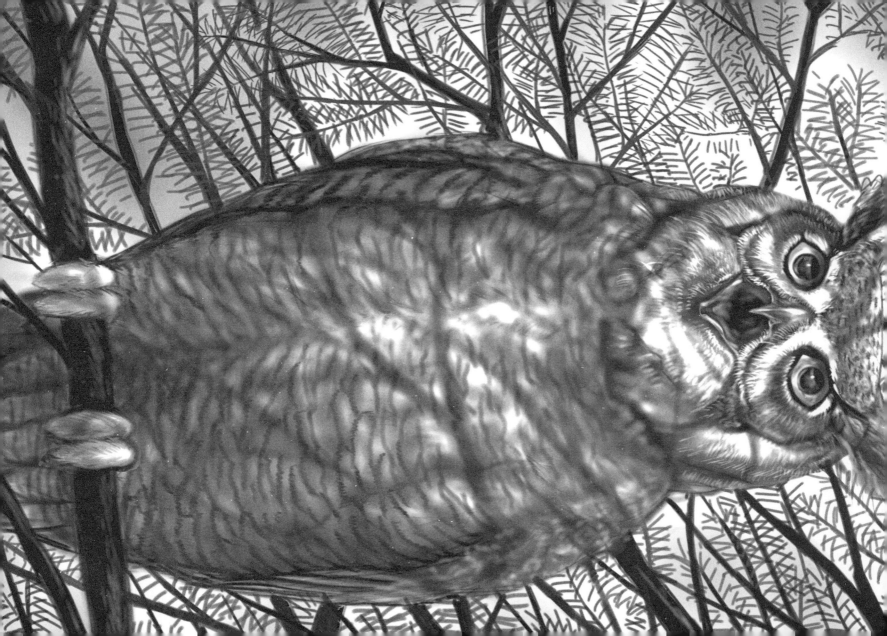

With the disappearance of the forest, all is changed.

—George Perkins Marsh, *Man and Nature* (1864)

Who? Who? Whoo? The annual interrogation started at 3 a.m. on a cold January morning. I was used to the interruption and, from the comfort of my bed, listened intently for any response. In the distance I heard a soft reply: *Hoo! Hoo! Hooo!* I would have preferred the harmony of two singers rather than the long-distance debate that was just beginning. This conversation—a typical territorial encounter between the two male great horned owls that divvy up my neighborhood—was going to last for a while. These owls, each of which stands nearly two feet tall, get an early start on spring by reasserting their property boundaries and courting in the heart of winter. As far as scientists know, they spend their entire adult lives in the same area, with their life-long mates. If the male in my yard is able to keep the neighboring owls away, then his partner might spruce up the old crow nest, high in the fir trees, and lay a clutch of two white eggs. She would then incubate the eggs for a month and brood the young owlets. As long as their efforts succeed, both members of the pair would hunt the neighborhood for mice, squirrels, and other small

Facing page: Great horned owl

animals to feed their growing offspring, who won't begin to fly from the nest until they are nearly two months old. As the parents work to usher in the next generation, my woods will quiet. The busy parents will softly hiss, meow, and coo, often composing duets. But then the real noise begins, for fledgling owls screech like frightened children all night long.

As much as I tire of night screams of young owls, I was thrilled to learn that two pairs of predators as formidable as great horned owls could find enough food and shelter to live among my family and our neighbors. Their presence would directly benefit our garden by keeping the nonnative eastern cottontail rabbits on edge. What was more, I knew that the owls were only one of many top predators that kept an eye on the streets and yards where I lived. As the owls hunted the night, Cooper's hawks, red-tailed hawks, and bald eagles hunted the day. There were smaller owls as well—saw-whet and screech—that hunted mice and large moths. Some predators that hunted here in the past were gone. At least at the moment there are no pumas, grizzlies, or wolves. But coyotes, black bears, and bobcats have all put in appearances. And this is my yard, not the wilderness!

My home provides shelter for more than a pair of large and powerful owls. The forested back acre I own provides a canopy, native groundcover, and brush piles adequate for two pairs of Pacific wrens and a pair each of spotted towhees, Pacific-slope flycatchers, and western tanagers. Three species of salamanders and two species of frogs seek shelter in the duff and deadwood that enriches my soil. Garter snakes hunt these amphibians and the slugs that are famously abundant in our cool, wet climate. The trees that live and die here house and feed a nice assortment of woodpeckers, chickadees, creepers, swallows, bats, and nuthatches. Fringing shrubs supplement my bird feeders with native berries, nectar, nuts, and bugs for busy grosbeaks, thrushes, juncos, and sparrows. The more I look and listen, the more my yard reveals. Though a key financial investment and a personal haven for my family and me, this property is much more than a mere commodity. As Aldo Leopold, the father of wildlife science, articulated so eloquently, my land is a community.

On a recent airplane flight I saw cities and their suburbs from a wider perspective. As we climbed out of Kansas City, Missouri, skyscrapers towered above the eastern deciduous forest. Below me the forest had been tamed, but not demolished. Dense stringers of trees connected seamlessly with city parks, streets, and subdivisions to outline big, lazy rivers. I saw an urban forest as a green quilt that sheltered two million people from the city's cold concrete and steel.

As we continued west, I could feel the exhaustion in the land below me. In contrast to the urban forest, the mosaic of land I viewed had been forced to conform to our mechanized world. Square ponds, rectangular fields, and crop circles defined this part of Earth. Massive irrigation projects held back and tamed the greatest of western rivers. Our replumbing of these aquifers endangers fish but enables crops to flourish where they could not naturally. The energy that once carved great canyons now energizes a power grid that cuts at right angles across the landscape and converges on cities and farms. Prairie and sagebrush have been pulled, scraped, and burned from the arid lands of western Kansas, Nebraska, and eastern Washington. Dark soils were fully plowed, planted, and starting to develop a ragged beard of green. These actions have made refugees of our native grouse, larks, sparrows, and buntings.

The depression I felt from seeing wall-to-wall agriculture eased as we met the jumble of western mountains. A lightly falling snow added mystery to a wild landscape. Most valleys were farmed or settled. And I could see where patches of timber had been downed and fossil fuels exploited. But there was

Earth at night (1994–1995 composite image, provided by Christopher Elvidge)

also a huge expanse of wilderness that allowed me to dream. As we passed over Yellowstone Lake, I forgot about people and wondered what the wolves were doing. I took comfort in knowing that below me many of the animals that avoid subdivisions and farms have some space.

My flight from Kansas City to Seattle reflected national and global trends. Throughout western Europe, eastern North America, and much of Asia our cities and suburbs dominate the land, literally lighting the night sky. But worldwide, less than 1 percent of Earth's soil surface is covered by urban development. In the United States this statistic is a bit higher—3 percent to 4 percent of the nation is a city, town, or village. The reach of cities is much greater. In 2010, one-quarter of the conterminous United States was designated as "metropolitan," because these areas held a city of at least fifty thousand people. Another 20 percent were deemed "micropolitan" on account of their economic and social connections to cities of between ten thousand and fifty thousand people. Our proximity to cities gentrifies nearly half of the U.S. national land base with well-tended lawns and gardens, public parks and recreation areas, and the infrastructure of transportation, communication, and energy. Worldwide, our coasts are densely settled, but even here, tree cover can be substantial. Currently, one-quarter to one-third of most urban regions around the world are covered with a canopy of trees or other natural greenery. Metropolitan areas in the United States contain an estimated seventy-five billion trees that shade a third of the area. These "lungs of the city," which produce oxygen and clean our air while providing essential habitat for other species, will be at risk in the future.

Agricultural lands fill in around our settlements and carpet much of eastern Europe, midwestern North America, Central America, and South America. One-third of Earth's land, and a full 40 percent of the United States, is farmed—far more than we occupy with our villages, towns, and cities. Parts of Africa, Australia, South America, and the Arctic remain wild and sparsely settled. Antarctica is a wilderness, but today this is the exception.

Most demographers expect our growing population to lead to more cities, especially moderately sized ones, widely distributed across the planet. In the United States, a million more acres become urban each year. As Harvard ecologist Richard Forman puts it, an urban tsunami is on the horizon. For many, the flood has already hit and the swimming is tough; one-third of all urban residents live in slums. When did this vast urbanization happen?

Five to six thousand years ago our ancestors in Mesopotamia and Syria created the world's first cities. As early populations rose and fell with environmental

Early Mesopotamian city

change, war, and disease, early agrarian settlements ebbed and flowed. At first, the majority of people lived outside of these creations, but as civilizations grew, so did the propensity of citizens to leave the dangers of the country for the comforts of the city. Large settlements took shape about three thousand years ago. Babylon, for example, in what is now Iraq, sheltered two hundred thousand. Between one thousand and two thousand years ago large cities were rare and centered in rich agricultural regions such as the Nile and Yangtze River valleys (Nanjing, China, may have been home to one million residents) and the Basin of Mexico (two hundred thousand lived in Tenochtitlán-Tlatelolco). Metropolises that we would recognize today are more recent inventions; three centuries ago only Constantinople, Edo (the former name of Tokyo), Peking, London, and Paris housed more than five hundred thousand people, and only thirty-four cities were home to more than one hundred thousand. The lure of cities proved irresistible, so appealing in fact that in 2008, for the first time in our history, more of us lived in cities than outside of them. Each year seventy-two million more people flock to cities. By 2050 more than two-thirds of all humans are expected to live in cities.

The city itself is only a part of the urban niche we have constructed. Like a galaxy, moderately large cities spin off suburbs, small edge cities, and a distant fringe of settlement called "exurbia"—"a semirural area beyond suburbia yet within its shadow." This gradient of urbanization from a densely populated urban core to a lightly peopled exurban fringe is dependent economically and socially on the commerce and culture of the city. The human census acknowledges this connection as it classifies Earth's residents. In the United States, for example, if we live in a city (areas with more than one thousand people per square mile) or in the associated less dense areas (at least five hundred people per square mile) connected to these cities, we are called "urban." In the 2010 census, as a resident of exurban Seattle, I became one of the urban people because the population density in my neighborhood crept over the lower critical threshold. Although my neighbors and I live on an average of two acres each—one hundred times more land than the residents average in the world's most densely settled city, Dhaka, Bangladesh—we are all urbanites. Our urban niche now includes commercial, industrial, and residential lands as well as their interstices of protected reserves, green recreational areas, and waterways.

It staggers me to contemplate the implications of our new lifestyle. We now live predominantly in a niche that was unknown only six thousand years ago! Generations of city people no longer have the interest to live in the country and may truly struggle to survive doing so. The social customs, diet, climate, modes of communication, and transportation in the city would be as foreign to our ancient ancestors as theirs would be to us. We have evolved into a new ecological role with cultural barriers to our rural legacy. Evolutionary biologists might consider us well along the process called "anagenesis"—the evolution of a new species from its ancestors that results from the gradual accumulation of isolating differences over time. Certainly we are already culturally distinct from our ancestors, but most biologists would not consider this adequate to proclaim that we are truly a species apart from ancestral *Homo sapiens*. They would require more lasting distinctions that make our DNA incompatible with that of our ancestors.

I have no doubt that our new niche and culture are, in fact, promoting genetic changes. As we prefer mates with new physical or cultural features, challenge our brains with the demands of new technologies, expose our metabolic processes to new diets, and succumb to new mortality agents, we expose our populations to new selective pressures that are capable of causing genetic, evolutionary change. Old foes such as dim eyesight, dull hearing, disease, and physical deformities no longer doom us to a short life with a reduced chance of reproduction. In their place are new hazards of urban life: fatty foods, carcinogens, inactivity, automobiles, and stress. And therefore, it is only a matter of time (and I doubt very much time) before our new lifestyle leads to speciation. Thus far, only about 250 generations of humans have been exposed to the rigors of urban life—an icing on the cake of our evolutionary history, as eminent evolutionary biologist Jerry Coyne points out. But I contend that this icing is an influential force significantly distinct from our earlier 300,000 generations of hunting and gathering that is able to shift our evolutionary trajectory. The day is near when we are no longer *Homo sapiens* (wise person) but instead something entirely new. Perhaps we are already *Homo urbanus* (city person).

As urban people we live in a time when everything changes rapidly. Gone is the Holocene, the geologic period that succeeded the Pleistocene, or ice age, twelve thousand years ago. Then, the climate was stable and life changed slowly. We now live in the Anthropocene—a period of chaotic change initiated by humans several millennia ago. Our fellow creatures quickly adapt, evolve, or die out. Like the great ice sheets that covered much of Earth during the Pleistocene, concrete and lawn now creep from our cities into the deep recesses of nature. How will these changes affect the community of life to which we are ultimately linked? Listening to the owls and others that live among us, we might hear the answer.

In visiting vast, primitive, far-off woods one naturally expects to find something rare and precious, or something entirely new, but it commonly happens that one is disappointed. . . . The birds for the most part prefer the vicinity of settlements and clearings, and it was at such places that I saw the greatest number and variety.

—John Burroughs, *Wake-Robin* (1871)

I'm not really a city person. I spent my formative years in small towns and faraway places in Kansas, Montana, Arizona, and Maine. As a research biologist, I spend a lot of time in the woods watching birds, not people. So when in 1997 I took a job in Seattle, Washington, I was quickly out of my element. My family settled outside the city, where trees cushion the view and neighborhoods include extensive natural greenways. But as I commuted to the city each day, I was struck by the extreme ways we engineer cities and how this engineering alters fundamental ecological processes. Just consider water. In a forest, rain is intercepted by leaves or soaks into the ground as it slowly carries important nutrients downhill to lakes, rivers, and the sea. When buildings and pavement replace forest, rain rushes to rivers in flash floods loaded with

Facing page: Anna's hummingbird

11

sediment and pollutants. When it suits our needs, we pump the river uphill or force it to flow, piped and underground, out of the sun's nurturing rays.

No wonder many of my colleagues consider cities unmitigated ecological disasters. Not only do we push around urban rivers, but we also construct buildings that funnel winds through urban canyons and foster industrial enterprises that heat the land. We are so noisy birds can't hear one another. Paving what was paradise removes and fractures habitat that other species absolutely require. Traditional ecologists have shown us the power we wield and the danger it poses to many other organisms on Earth. Hit songs such as "Big Yellow Taxi" by Joni Mitchell and "The Last Resort" by the Eagles have popularized the notion.

I see these changes and feel their effects every day as I ride a bus from the exurbs to the city. I shed my coat as I enter the urban heat island, pass through planted woodlands of trees from around the world, and cross a bridge that spans a ditch dug to connect the once-inland Lake Washington to the Pacific Ocean. But I also encounter a rich diversity of birds. Majestic bald eagles hunt from streetlamps. Once-endangered peregrine falcons nest under steel bridges or atop skyscrapers. Cormorants decorate the cottonwood trees as they strike a silent pose to dry outspread wings in infrequent sun breaks. Crows, gulls, and pigeons dine on our leftovers. Parties of bushtits, flocks of robins, and mixed groups of nuthatches, chickadees, and kinglets enliven the shrubs and lawns on which I walk. A paradox eats at my subconscious. Everything I have learned as a conservation biologist tells me cities are bad for biodiversity—the sum total of life in an area—yet the feathered collective I encounter seems wholly unconvinced.

Maybe my problem is that I live in Seattle. After all, this "Emerald City" nestled in wild country is young. European settlement did not begin until 1851, and even today buildings are interspersed among trees, rivers, lakes, and the Puget Sound of the Pacific Ocean. Forests of towering native Douglas-fir, western hemlock, western red cedar, red alder, big-leaf maple, and black cottonwood mix with street trees to provide a green canopy over nearly one-quarter of the city. The forests of Seattle and its suburbs now embrace 141 species of trees, including 30 native species and ornamentals from North and South America, Europe, Asia, and Africa. Some are problematic invaders, but in total they provide a diverse menu of foods and nesting and roosting sites for birds. Birds flock to my green city. Surely they must shun the concrete elsewhere.

A few trips persuade me that this is not the case.

I travel with binoculars. I'm not out to tally an impressive state-, country-, year-, life-, or even dream-list. I simply enjoy watching birds in new places, and this predilection affords me an opportunity to question the generality of my Seattle observations. As I follow my avian ambassadors into the natural side of our built world, they take me into trashy gullies, streams that run over partially submerged shopping carts, and dangerous neighborhoods where a visitor who cares about safety should not tread. Amid warnings from concerned residents, I manage to confirm around the world what I've seen in Seattle. The downtowns of cities in North and Central America, New Zealand, and Europe are rife with birds. My daily counts average twenty-three different species and range from a low of eleven in Auckland, New Zealand (where only four are natives), to a high of thirty-one species in St. Andrews, Scotland, where all but four are indigenous. In Berlin, Germany, I watch white-tailed sea eagles fish the Wannsee, just like Seattle's bald eagles plunge into Lake Washington. Dippers, chunky aquanauts that pump up and down on streamside perches, wade into icy streams in Ketchikan, Alaska, and St. Andrews to search under rocks for caddisfly and mayfly larvae. I recollect these and other colorful birds—red-billed gulls, long-tailed ducks, yellow-winged caciques, rufous-naped wrens, red-bellied woodpeckers, crimson-fronted parakeets, ferruginous pygmy-owls, and oystercatchers—from my urban visits.

I see a few of the same birds in nearly every one of the ten cities I visit. Five are particularly cosmopolitan. All are the result of human actions. From Europe, thanks to the desires of acclimatization societies, most cities now harbor house sparrows and European starlings. From North America, the interests of duck fanciers and hunters provided mallards and Canada geese to the urban world. The rock pigeon (or common street pigeon), originally domesticated from wild Mesopotamian stocks five thousand years ago, may be the world's most familiar city bird. These "fab five" exemplify what conservation biologists call "biotic homogenization," an increasing similarity in the flora and fauna of distant lands once isolated by geography but now joined by the industry of human transportation. I worry about this result but take some comfort that the fab five are the exception; most striking to me is the regional distinctiveness of each city's birds.

In the downtown core of my ten cities I rack up 151 unique bird species. For every four birds I see, I find three only in a single city. Homogenization is barely perceptible. I uncover unique representatives of the waterfowl, raptor (hawk and owl), corvid (jay and crow), dove, finch, woodpecker, and sparrow tribes in each location. For instance, I see gulls of the ring-billed, black-headed, red-billed, black-backed, glaucous-winged, Thayer's, and Franklin's variety. Of the black-backed type alone, I spy three species, each in a different city. Woodpeckers are nearly as diverse; black, great-spotted, downy, hairy, Hoffman's, and red-bellied are on my list. Double-crested cormorants occur in New York and Ketchikan, but in the latter city I also find pelagic cormorants, whereas I find the neotropical variety and the shag in Zihuatanejo, Mexico, and St. Andrews, respectively. My ten cities harbor nine species of corvids, eight species of titmice ("chickadees" to Americans), six ducks, six doves, four herons, four raptors, three swallows, and three hummingbirds. The only partridge is the ring-necked pheasant, an alien species introduced from Asia to Scotland to appease hunters.

My surprise at the bounty of city birds is equally matched by its comparability to that of nearby wildlands. My count in Ketchikan is almost double the next day's count along the Naha River—a remote wilderness fifty miles away

Finding Subirdia

The shoreline in Ketchikan, Alaska, teems with rare—harlequin duck—and common—mallard and Canada goose—waterfowl.

in the Tongass National Forest that required powerboat, kayak, and hiking to attain. But the real stunner comes from a comparison of two signature parks in the United States. Yellowstone National Park is a wild gem where wolves and grizzlies sit atop a food chain that characterizes a true primeval ecosystem. The park is a whopping 2.2 million protected acres within a wild ecosystem of nearly 20 million acres. Each March I spend a few days on the park's northern range. In 2013, after counting twenty-six bird species in Yellowstone, I flew to the fourth largest megacity in the world, New York. There, in the heart of Manhattan, walking distance from Broadway, Fifth Avenue, and Wall Street, is a "wild" park of a completely different order, Central Park. Thanks to the foresight of landscape architect Frederick Law Olmstead, eight-hundred-acre Central Park brings nature to the city.

As I bird north along Sixth Avenue, I see only house sparrows, European starlings, and rock pigeons. Then I enter Central Park and quickly find

House sparrows, European starlings, and rock pigeons feed on grain spilled by the carriage horses of Central Park in New York City.

mallards and Canada geese. I expect to see little else. But *fugetaboutit!* I can barely keep up with the rush of cardinals, blue jays, white-throated sparrows, black-crested titmice, mourning doves, black-crowned night herons, wood ducks, cormorants, red-tailed and Cooper's hawks, crows, blackbirds, and three varieties of woodpeckers. I leave after two mornings of birding, having seen more species—thirty-one—than I saw in four days of traveling through Yellowstone! Perhaps I should have expected more species in the rich deciduous forest of New York than in the high montane west, but I figured my longer visit to Yellowstone would compensate for regional differences. Historical records since the late 1800s suggest that about the same number of bird species—two hundred—are regularly encountered in both parks. From a bird's perspective, a large park created by human hands or by nature is not all that different.

With my graduate students I have counted birds from Seattle's urban core to its fringing forests nearly every spring and summer morning for the past decade. Most of the forests we studied held few giant and ancient trees—loggers cut them down a century ago—but substantial evergreens typically around one hundred years old were plentiful. Here, with standardized efforts, we can truly gauge the effect of urbanization on the variety of birds. When we stood quietly in the industrial downtown of the city for a ten-minute count, we would tally ten to fifteen species. That might not sound like many, but to be counted the birds must be seen or heard nearby—that is, within about 150 feet of us. Pigeons, crows, house finches, and house sparrows were common, as they are in any western North American city. Occasionally, an Anna's hummingbird buzzed us. This little jewel has taken advantage of warming climates and feeders to move north and invade Seattle, even living in our cold mist during the winter months. When we went far from the city to places our society has reserved for nature and clean water, we found a few more species, usually twenty birds that are typical of northwestern forests such as Townsend's, Wilson's, and black-throated gray warblers; chestnut-backed chickadees; Swainson's thrushes; pileated woodpeckers; and Pacific wrens.

We expected the suburbs between the city center and the forested reserves to support an intermediate number of species, but when we listened as these neighborhoods awoke each morning, we were astonished! We did not hear the *hoot* of the endangered spotted owl or the *keer* of the rare marbled murrelet, but the dawn chorus of thrushes, tanagers, wrens, towhees, finches, crows, and woodpeckers quickly cleared the stiffest cobwebs from our brains. Here we often tallied thirty or more species in a ten-minute count. We found those birds from the industrial city mixed with some of those from the protected forest. And we encountered a whole new set of birds that use more open country such as violet-green swallows, willow flycatchers, killdeer, orange-crowned warblers, American goldfinches, and Bewick's wrens. Ethereal music typical of open headlands or the tundra serenaded us, courtesy of male white-crowned sparrows.

Compiling standard bird surveys from more than one hundred locations in and around Seattle revealed to us a consistent, but unexpected, relationship between the intensity of development and bird diversity. The greatest diversity was not in the most forested setting. Instead, bird diversity rose quickly from the city center to the suburbs and then dropped again in the extensive forest that eases Seattle into the high Cascades. We had discovered *subirdia*.

Now I was really perplexed. Suburbs, you see, are not only rich in birds, but they are the preferred habitat of my fellow Americans. First designed in the mid-1800s outside New York City and Chicago, suburbs now house more than 40 percent of all U.S. residents. Our suburban lifestyles were made possible by advances in transportation and agriculture. Level and cleared farmland around cities was available for settlement through the 1900s as farmers were able to increase yields fivefold and efficiently ship goods from distant farms to urban markets. Trains, streetcars, and eventually private automobiles allowed workers to commute between home and office. This mobility and our quest for the American Dream—which includes private homeownership—pushed suburbs far from the city. The search for quiet living often caused new suburban developments to leapfrog over existing ones, producing urban sprawl. Sprawl consumes land and fuel, requires networks of roads, taxes those who live in the city, and stresses out those who commute. Some planners and architects see the potential of this disorganized landscape, but its lack of cultural identity has caused many to describe it as the "geography of nowhere." Sprawl is a factor in the endangerment of many species, including some birds. How could suburbia also be a mecca for birds?

Human neighborhoods are good for birds because they offer a wide range of habitats in a small area. Lawns and trees are jumbled into savannahs, fields, and woodlots. Engineers provide new features such as small ponds that retain runoff from the many sealed surfaces. Where different habitats touch, they produce rich edges that offer access to many resources, such as nuts from trees, seeds from annual weeds, and insects from ponds. The diversity of plants found together in urban settings is simply incredible. In a single garden in Leicester, England, 422 plant species were recorded. A broad survey of sixty-one urban yards throughout the United Kingdom revealed 1,166 vascular plants, 80 species of lichens, and 68 bryophytes (mosses and their relatives). Suburbs are compact microcosms of the plant world. While not as diverse as tropical rainforests—two acres of Yasuni National Park in Ecuador contain 655 tree species—they are more diverse than most of Earth's natural forests.

Plant diversity begets a diversity of bird foods. Birds can find, of course, a cornucopia of nuts, berries, and fruits, but there is more. The specious garden in Leicester was infested with 1,602 species of insects and 121 other species of invertebrates. A survey of twenty-one gardens in the United States turned up 110 species of bees. Nearly every breeding bird feeds its nestlings

Subirdia (darkened areas) fringes cities throughout the United States (image from 2006 National Land Cover Database classified by Fry et al. 2011).

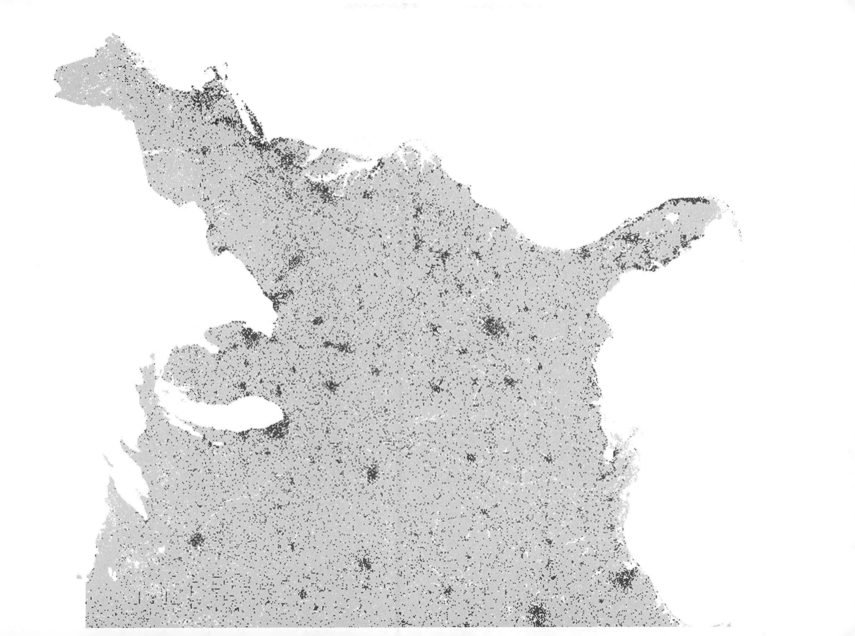

a steady diet of insects. Residents also supplement the diverse urban forest with generous offerings of birdseed.

Our discovery of subirdia in Seattle is not unique. Throughout Britain, in deciduous woodlands of California and Ohio, grasslands of Arizona, forests of Japan, and shrublands of Australia moderate levels of urbanization also provide an abundance of various resources that increases the number of bird species beyond that found in either wilder or more densely populated settings. However, this is not always the case. The peak in bird diversity occurs where the creative hand of urbanization surpasses the destructive hand. In temperate locales, the creation of habitat diversity allows more birds to colonize suburbs than are extinguished by the destruction of unique habitats required by shy and intolerant species. In tropical settings, where magnificent rainforests rife with birds are converted to suburbs, bird diversity drops. And it plummets as one moves ever closer to the tropical city's center.

I awoke on my first morning in Alajuela, Costa Rica, to the noise of motorbikes and parrots. I bolted from bed to find scattered pairs of crimson-fronted parakeets perched among the trees, electrical wires, and buildings. Like footlong emerald ornaments with pointed tails, red freckles, and golden underwings, the birds decorated the central plaza. They wheeled about, grooming each other and mixing freely with street pigeons. Such tropical cities hold avian riches, but their splendor is deceiving. Jeff Norris knows this deception. He earned his Ph.D. by counting birds along six separate urban-to-wildland gradients in Central America. These gradients each included the full range of developments that occur between a city and a nearby forest reserve, such as a national park. Jeff discovered on both the Atlantic and Pacific sides of Costa Rica that bird diversity peaks along the edges and deep into the forest reserves that compose nearly 30 percent of this Central American nation. He counted up to 116 species in the forest, but no more than 53 in the suburbs. Counts of 20 to 30 bird species were more typical of the city center. Other scientists have found similar results in tropical Mexico, Ecuador, and Singapore. With so much to lose, tropical subdivisions will need to offer birds much more than they currently do before they too can become subirdia.

Attaining and maintaining a rich avifauna are worthy goals for urban areas. Birds, even nonnative parrots that have found their way into many temperate cities, inspire people to care about other species. But enthusiasm for diversity carries responsibility. It requires us to track worldwide changes and guard against the loss of a region's distinctive life. House sparrows, European starlings, mallards, Canada geese, and rock pigeons—that fab five—live in nearly every city around the world. Within a continent, the number of birds held in common among cities is greater than the tally of cosmopolitans in less developed settings, such as forests or grasslands. This is the start of the homogenization process. By creating cities that birds view as similar—offering the same foods, plants, and arrangements of buildings and parks—we are starting to erase the diversity that we can celebrate. So, while we cheer the parakeets of Telegraph Hill, we should also recognize that they reduce the uniqueness of San Francisco by increasing its similarity to tropical cities a hemisphere away.

To be a bird living among *Homo urbanus* is no small feat. It requires a nimble nature with the ability to colonize new opportunities and persist when the known world suddenly is transformed. As we celebrate urban diversity, let's now take a close look at the dynamics of colonization and extinction. The forest just beyond Seattle has changed in the short time I've studied its birds. By going among the bulldozers, asphalt pavers, and nail gunners, my students and I have watched many birds adjust and their populations soar. A few others are less fortunate. As we consider how Pacific wrens face up to the big and hostile Bewick's wrens, or what happens when traveling juncos, robins, and sparrows suddenly discover the riches of someone's yard, we can appreciate how subirdia develops.

THREE. *A Child's Question*

In 1898 the famous naturalist W. H. Hudson published a detailed survey of the birds, which, though it recorded some recent colonists, gloomily forecast the steady impoverishment of the avifauna of [inner London]. The reality has been very different.

—Stanley Cramp (1980)

What's going to happen to the animals?

It seemed to be a simple question, but it bugged me for forty years. In the early 1970s, when I was twelve or thirteen, I asked my dad about some survey stakes that sprouted up in the field and woods at the end of our street. My buddies and I used this patch of Kansas soybeans and the forest it fronted as our personal wilderness. We filled our packs and walked to the woods each summer to explore and escape. We caught frogs, snakes, and turtles. We camped, shot grasshoppers with our BB-guns, and perfected our prowess with axe and hammer. We relaxed in tree houses and listened to the sounds of the deciduous forest and small streams. The markers we found worried us. Would "our" forest be bulldozed? What would happen to the deer we tracked, the birds we pursued, and the adventure we enjoyed when our piece of paradise was paved?

Facing page: Bewick's and Pacific wrens

25

Dad said the animals would move to a new place. His life in the military and recent conversion to a career in real estate lent more than his usual authority to the response. In fact, the stakes marked a new subdivision he and his partners were developing. But I think we both knew the answers weren't that simple. The animals couldn't just resettle elsewhere. There were other animals in the places that might harbor the refugees. The wildlife would either have to adjust to their new human neighbors or perish.

Four decades later, it was time to find out which alternative had played out. The place I knew as a kid was transformed into the one that now stood before me.

At dawn, on an April day in 2012, I parked the rental car in the old neighborhood where the bean field used to stand and headed into my childhood woods. Spring seemed to be in a hurry to erase a winter that never really happened. The once-mysterious small creek still cut through rusty and white limestone as it headed north toward Baldwin Creek and the muddy Kansas River. A mix of native and introduced vegetation along the waterway provided an occasional buffer from new roads and homes. Exotic honeysuckle bushes, something unknown in my childhood, were everywhere. (Scientists studying a site in Ohio have discovered that these invaders provide birds with cover and nesting places but that nests in honeysuckle often suffer high rates of predation.) Today, along with blooming native black locust, they perfumed the air. Vines of Virginia creeper provided a tropical cloak to the locust, Osage orange, redbud, catalpa, and feral fruit trees that made up the forest. If I ignored the honeysuckle, the types of trees and bushes I remembered still remained. But what would have been an all-day adventure in my youth was now a thirty-minute stroll on paved sidewalks and a mowed powerline right-of-way. The forest that had beckoned young explorers was mostly reduced to a riparian amenity for huge homes. Massive lawns fronting streets that boasted of a wilder past isolated the trees that remained: Tallgrass Drive, Timber

Court, and Cattleman Trail. A fragment of its earlier extent, my forest was patchy and invaded, but green enough to lure me closer.

As I birded my way through the subdivision, I was drawn back in time: Brown thrashers still sing the repertoires they pilfer from others and then, as if embarrassed by mimicry, dart into the undergrowth. Pairs of cardinals are everywhere, nearly one duo per garden. No longer the shy resident of distant woods as they were a century and a half ago, they seem to outnumber even the robins. There are plenty of native cavity nesters, including white-breasted nuthatches that I see carrying food to nestlings, tufted titmice, black-capped chickadees, downy woodpeckers, and red-bellied woodpeckers. There are no red-headed woodpeckers; this once-common acorn specialist is now rare everywhere. A red-tailed hawk uses the power poles to hunt the grass for voles and mice; I suspect a nest nearby. I tally eighteen species in a morning survey. Most of what I count are birds that live here year-round; it is too early for many migrants to have returned. Yet I am surprised to not have encountered an early-arriving warbler such as the Louisiana waterthrush; the only warbler I tally is the common resident yellow-rumped warbler. Perhaps these woods are too dissected, sprayed with insecticides, and groomed to provide insects and solitude for migrants facing the unusual dryness of a mild winter and early spring bloom.

People who now live in my childhood wilderness seem busy and intrigued with nature. I see lots of bird feeders and no loose cats. But nobody seems to veer from the streets or walkways. There are no trails in the grass, and none penetrate the woods. I find only one old tree house—near the place I used long ago with my friends. I'm glad to know some kids have been here. But as I scramble through the redbud to get a closer look, wondering whether this fort could be ours, I see little sign of recent use. The sprawling tree supporting this hideout has fallen, and the carpet, plastic tarps, and wood are scattered about. Maybe some of the wood was recycled from our efforts, but this was not our hideout. It seems impossible that my sense of this place is forty years old. The taming of the land seems to have tamed the humans.

A moist clearing in the forest has a few deer tracks. So, individuals of this species adapted! I am relieved that my dad's actions did not extinguish these large mammals that so thrilled me years ago. A rich understory without obvious signs of browsing and a lack of caging around neighborhood plants suggest they are not overly abundant. Perhaps they did not move as Dad thought, but neither did they go extinct. A few simply hung on to the little forest and field that the development retained. The chittering of American goldfinches raises my spirits as I recall with guilt shooting one as a kid in these same woods. I'm reassured that they are still here and abundant. Specialized finch feeders and expensive seed feed these gilded treasures. I'd like to know whether tanagers and cuckoos will return later in the spring, but I am not complaining. I expected much worse. Only two of the bird species I encounter, the house sparrow and European starling, are not native. The native species I find are much more interesting and include predators, cavity nesters, and those that prowl bushy thickets. Ground nesters are rare. Of the birds I find, only the brown thrasher is known to nest on the ground, and even it is not obligated to do so. I walk back to the car, lighter on my feet than when I started, thinking of past adventures. Despite losing a forest, Lawrence, Kansas, has retained a woodland that hosts native birds and deer. I hope their tracks and songs will lure another kid from the easy path into the cool woods where a tree house can be resurrected and a love of nature kindled.

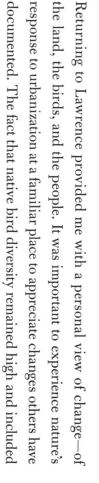

Returning to Lawrence provided me with a personal view of change—of the land, the birds, and the people. It was important to experience nature's response to urbanization at a familiar place to appreciate changes others have documented. The fact that native bird diversity remained high and included sensitive and adaptable species, but also reflected losses of favorites such as the red-headed woodpecker, was to me personally relevant. Such relevance was, however, not scientific. I had no benchmark of a rigorous bird survey from my childhood woods. The current assembly of birds seemed okay, but what was really lost or gained over those forty years? Scientific retrospectives in Virginia, Australia, and the United Kingdom suggest some answers.

For centuries, a mature, deciduous forest cloaked the southern shore of Lake Barcroft, a reservoir serving Alexandria, Virginia. Situated a mere eight miles from the White House and its First Family, this region of northern Virginia (Fairfax County) was home to the newly minted U.S. Fish and Wildlife Service's ornithologist, John Aldrich. Aldrich was a distinguished researcher whose career would span fifty years and was bookended by bird counts around Lake Barcroft. Shortly after moving to Alexandria in 1941, Aldrich employed the precise and time-consuming methodology of his graduate mentor, S. Charles Kendeigh, to map the territories of all breeding songbirds in a ninety-five-acre forest fronting the lake. The young ornithologist delimited 187 territories of twenty-three species. Most common were the obligate, deep forest species. All but 39 territories were owned by pairs of red-eyed vireo, ovenbird, wood thrush, scarlet tanager, hooded warbler, Acadian flycatcher, and eastern wood pewee. And then development happened.

Beginning in 1950 Aldrich's study area was transformed from forest to suburb as much of northern Virginia rapidly urbanized. The urban tsunami washed over Lake Barcroft, and in its wake the ninety-five-acre wood splintered into a mixture of forest remnants and single-family residences, one of which was occupied by the Aldrich family in 1959. To Aldrich, science was now getting personal.

Ornithologists are obsessed with the birds around their homes. Aldrich watched his birds whenever he could and slowly accumulated a lifetime's perspective on and passion for Lake Barcroft's avian world. In 1979, retired from federal service but not yet released from the lure of science, he remapped the bird territories in his neighborhood. Gone were the most common birds of the 1940s. There were no pairs of red-eyed vireos, ovenbirds, scarlet tanagers, hooded warblers, Acadian flycatchers, or wood pewees. In their places were cardinals, mockingbirds, song sparrows, blue jays, starlings, gray catbirds, American robins, and house sparrows. In total, the density of birds increased (from 187 to 263 territories), and the number of species rose by six (from twenty-three to twenty-nine). The avian community grew as it was transformed; eleven species were lost while seventeen were gained.

A similar story played out across the Atlantic Ocean in London, England. There, in the central forty square miles of the world's twenty-third largest city, ornithologists surveyed birds in 1900 and again in 1975. Rather than documenting a startling decline in the variety of birds, they were surprised to find that nearly all birds known from 1900 remained in London seventy-five years later; only five species were lost. To offset these losses, twenty new species colonized the city, substantially increasing London's bird riches from twenty-five to forty species. And the number is increasing still; the 2008–2012 London bird atlas project recorded sixty species! Reduced pollution, improvements in public attitudes and actions on behalf of birds, and conservation efforts to restore bird numbers and embellish bird habitat in city parks contributed to the renewal of London's birdlife.

In Perth, Australia, birds have not fared quite as well. In the central but large and wild Kings Park, ten species known in 1937 were lost by 1986. Most species survived within the park and held their own for the five decades spanned by the study. But only three new birds colonized the park during this time, so the number of species dropped from thirty-nine to thirty-two. Trampling of ground vegetation and enhancement of nectar, fruit, and water in suburban gardens likely caused the changes in Perth. Birds that pounce on ground insects such as the sacred kingfisher and scarlet robin were among those lost, whereas species that fed above forest openings and on ornamental fruits—such as the welcome swallow and rainbow lorikeet—were gained.

Australian researchers suggested that increasing native vegetation, especially on the ground and among the shrubs, may reverse Perth's losses, much as was the case for London.

The long-term transitions at places like Lake Barcroft, inner London, and Kings Park reveal slight changes in the membership of bird communities and substantial changes in the abundance of the birds that remain throughout a period of urbanization. In all cases, the birds that were most common before the expansion of urban or suburban lands declined in numbers, while a set of those that were initially rare increased. In Virginia, for example, the wood thrush held thirty-one territories in 1942, but only two in 1979. In contrast, the number of cardinal pairs during this time increased from five to thirty-eight. Dr. Rob Blair, a pioneer in the study of urban birds and professor at the University of Minnesota, categorized species by such responses to urbanization as avoiders, exploiters, or adapters.

Avoiders are those species that are extinguished or decline precipitously over time as urbanization intensifies. In the eastern United States, some avoiders are the forest specialists such as the red-eyed vireo, ovenbird, and wood thrush. In western Australia avoiders are the ground pouncers. In London they are the cavity nesters such as the jackdaw and species with special needs such as the skylark, wryneck, and nightingale. What we do—pave, turf, light, pollute, fragment, disturb, and make noise—they cannot tolerate.

Adapters, on the other hand, are mostly native species that thrive on natural young, open, shrubby, and dissected native habitats. They find and adjust to these situations in our cities and towns, even if the natural habitat is only approximated by the built condition. Adapters are tramps easily dispersing across our world. If they were plants, we'd call them weeds. They live fast, prodigious lives and die young. Their natural strategy is to track environmental disturbance—they are the first to colonize lands perturbed by hurricanes, fires, tsunamis, or glaciers. Because our cities produce and maintain the sort of features that follow disturbance—grassy meadows, brushy slopes, or rocky

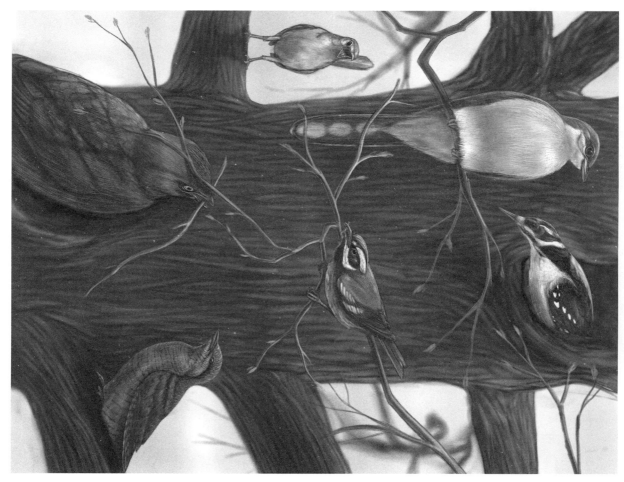

Avoiders (clockwise from upper left: yellow-billed cuckoo, hairy woodpecker, black-throated gray warbler, wryneck, jackdaw, nightingale)

strands—adapters naturally take to our settlements. Around the world, finches (cardinals, goldfinches, greenfinches), thrushes (robins, song thrushes, European blackbirds), sparrows (dunnocks, song sparrows, white-crowned sparrows), tits (great tits, blue tits, bush tits, black-capped chickadees), corvids (magpies, ravens, and some crows), and birds that sip nectar from opulent flowers (singing and brown honeyeaters, Anna's and cinnamon-tailed hummingbirds, rainbow lorikeets) are adapters.

Exploiters are species that thrive in our presence, often coevolving with humans and rarely occurring where people do not exist. Exploiters nest or roost on our homes, eat our waste, and walk our streets. They are among our most familiar birds and include the fab five cosmopolitans we met in Chapter 2. Other exploiters are American and hooded crows, ring-necked and monk parakeets, northern and tropical mockingbirds, black redstarts, barn swallows, barn owls, chimney swifts, house martins, house finches, and house wrens.

Identifying avoiders, adapters, and exploiters is possible by comparing bird communities in areas that have changed over time or by comparing several places that differ in the degree or age of settlement at one time. Unfortunately, these studies cannot tell us how or when the avoiders declined, or when and in what order the adapters or exploiters appeared. We cannot infer whether the gains and losses were sudden or gradual. Hidden are the possible interactions between colonists that may determine their tenure in our towns. Gaining such understanding requires a finer look at the appearance and disappearance of birds in neighborhoods over time. It was time to search for more survey stakes and carefully map the response of birds to the changes that were certain to follow.

Roarke Donnelly grew up outside of Chicago and moved to Seattle in 1997 to pursue his Ph.D. As he considered possible dissertation topics, he was struck

Adapters (clockwise from upper left: goshawk of European cities, white-crowned sparrow, northern cardinal, black-billed magpie, European blackbird, great tit)

Exploiters (clockwise from upper left: barn owl, house finch, barn swallow, black redstart, hooded crow, ring-necked parakeet)

by the responses of birds to urbanization. Needing a project that he could complete in three or four years, he undertook a survey of Seattle's birds, focusing on the influence of park size and subdivision configuration on avian abundance and diversity. But as Roarke counted, he also noticed signs announcing forthcoming development. In the lab, we talked about the changes to our childhood neighborhoods and considered the future of the Seattle area that seemed eager to trade trees for homes. We decided to begin a study of places destined for development. Surveys before construction would be useful for Roarke's study, but for me and a dozen other students who followed his lead, they would provide the baseline, replication, and controls needed to apply science to my childhood question of what's going to happen to the animals.

As the 1990s came to an end, we had preliminary counts of birds at five reserves, ten established neighborhoods, and eleven sites slated for development. I wrote grants to fund a legion of students and technicians to join me each spring and summer to count, map territories, catch, band, and find nests of the birds that used each of these sites. Sometimes we had to disguise ourselves as construction workers or prospective homebuyers to enter our study areas. Some developers were hesitant to let us watch birds as they cleared the land, created ponds, and built homes, but my students were undaunted. Residents also wondered what we were up to with binoculars on their street corners and back acreage. We explained our study to hundreds of people, including some who carried badges, Tasers, and guns. By 2010, we had counted birds during nearly six thousand standardized, ten-minute-long surveys on our twenty-six sites. That effort—during the wee hours of more than five hundred mornings—allowed us to tally more than fifty-five thousand individual birds of 111 species. Our sleep deprivation had produced a detailed picture of how birds in subirdia responded to the process of urbanization that we could compare with nearby forested reserves and previously settled neighborhoods.

The Avoiders: In our temperate, evergreen forests we found extinction to be rare, though a quarter of the small forest birds declined as the new subdivisions were constructed and inhabited by suburban people. Three species, the northern spotted owl, the yellow-billed cuckoo, and the marbled murrelet, which are hypersensitive to forest loss, had disappeared from our inhabited, lowland forests decades earlier. They were also the only extinctions recorded just north of the border in Vancouver, British Columbia. Ten of the forty-four species that we found before development were avoiders, and each declined as construction ensued. In contrast, the abundance of these species was stable in nearby forested reserves over the same period of time, and they were much more abundant in these forests than they were in the established neighborhoods we also surveyed. The recent and ongoing avoidance of development by the ten extant species we studied suggests that others may soon follow the owl, cuckoo, and murrelet. Certainly avoidance of built areas is an important reason why many of them are declining throughout western North America at rates of up to 2.5 percent per year.

We can rank our ten avoiders according to their sensitivity to development by considering the frequency with which they were locally eliminated. Atop that list is the Townsend's warbler, a stunning black-and-white denizen of the upper coniferous canopy who sports a face striped with black and yellow. While generally rare at our lower elevations, the wispy songs of this warbler were silenced in all forests we observed being converted to neighborhoods. Its close relative and resident of lowland maple forests, the black-throated gray warbler, also disappeared from two-thirds of the neighborhoods. Pacific wrens, mouselike birds that scurry among the ferns and perch on hundred-year-old stumps to pour forth sweet concertos, were uprooted from four developments and reduced from local prominence to a measly pair or two at three other sites. The remaining avoiders were all lost from at least one site, but hung on during the tumultuous period of forest conversion at a majority of the neighborhoods we studied.

The avoiders were migratory and year-round residents. Six of them—Pacific-slope flycatcher, Swainson's thrush, western tanager, and Townsend's, Wilson's, and black-throated gray warblers—were neotropical migrants.

These species breed in western Washington but overwinter in western Mexico, Central America, and South America. Many migrants are on the decline as they face challenges in two worlds as well as along the migratory route that connects them. They avoid development because it removes understory nesting cover used by the thrush, flycatcher, and Wilson's warbler and the upper canopy used by the other warblers. Similarly, resident Pacific wrens and golden-crowned kinglets require native groundcover and canopy cover, respectively, and so, too, avoid development. The last two avoiders are residents that require dead trees. The hairy woodpecker drills its nests and nighttime roosts in dead trees that average ten inches in diameter. The brown creeper is a tiny bird that hitches up tree trunks probing the bark with a long, narrow, decurved beak in search of spiders and insects. It builds a soft nest out of lichen, moss, fine rootlets, and hair behind flakes of peeling bark that develop in the first few years after a tree dies. Dead trees, or "snags," are critical resources for many birds, but construction crews and homeowners concerned with safety quickly remove them. That many small snags remain in our forests may explain why both creeper and woodpecker populations are faring better than the other development avoiders across the West.

The tenure of some avoiders within recent developments may be short-lived. This possibility haunted Cara Ianni, a young educator from the Seattle area who believed it was important to teach from experience. Seeking exposure to conservation biology, Cara began her graduate research in 2003 to determine how birds fared as Seattle's subdivisions aged. She surveyed birds in thirty-five neighborhoods that ranged from five to one hundred years old. She quickly saw that older neighborhoods lost their forest-dependent birds. Not a single Townsend's warbler or hairy woodpecker could be found in any of the seven neighborhoods that were seventy or more years old. Black-throated gray warblers, Swainson's thrushes, and Pacific-slope flycatchers were found in only one old site. Our other avoiders remained, but were rare, in Seattle's oldest neighborhoods.

Rufous hummingbird

The Exploiters: As the troubled avoiders winked out of new subdivisions, seven exploiters replaced them. Included were the fab five cosmopolitans—Canada goose, European starling, house sparrow, mallard, and rock pigeon—plus two locals, the American crow and house finch. Each species soared in abundance as soon as clearing began and increased twofold to thirtyfold during the next decade. All of these species are virtually nonexistent in forest reserves, and finches were rare at new neighborhoods immediately after clearing began. The others did not colonize until three to five years later. The ability of these species to capitalize on the lawns, lakes, and nesting niches created in built environments likely explains their regional success—all except the house finch and house sparrow are stable or increasing across western North America. Finches and sparrows may be declining in response to factors beyond settlements, such as the socially transmitted disease conjunctivitis that affects house finches and the loss of small-scale agriculture that is of critical importance to house sparrows.

The Adapters: Far and away the largest group of birds that responded to development were those that adapted to new opportunities. Twenty-six such species increased twofold- to one-hundred-fold during the creation of new subdivisions. All but three—black-headed grosbeak, chestnut-backed chicka-dee, and yellow-rumped warbler—were more abundant in established devel-opments than in forest reserves. About half of the adapters colonized or peaked in abundance midway through the development sequence, before avoiders were eliminated. They used environmental features created by clearing and construction such as the ponds and gravel pads defended by red-winged black-birds and killdeer, dead tree promontories commandeered by olive-sided fly-catchers, and weedy hillsides ruled by American goldfinches, Bewick's wrens, savannah sparrows, white-crowned sparrows, and orange-crowned warblers. Some of these species continue to persist in older neighborhoods, but only at very low densities.

Adapters included a mix of neotropical and shorter distance, regional mi-grants as well as residents. They nested on the ground, in shrubs, and in dead tree cavities; forest canopy species, as noted earlier, avoided development. Adapters ate what subdivisions offered—insects, seeds, fruits, and nectar. They searched for their meals on lawns, in forest openings, along marsh edges, and on bare, rocky ground. They were an extremely diverse lot of open-country birds plus a few birdseed addicts such as the black-capped and chestnut-backed chickadee.

The ephemeral nature of the habitats adapters used is reflected in their regional population trends. Most are declining along annual survey routes scat-tered throughout the western United States, some substantially so. Pine siskins, willow flycatchers, rufous hummingbirds, killdeer, and American goldfinches have declined annually since the late 1960s by an average of 2 percent or more on western Breeding Bird Survey routes. (The Breeding Bird Survey is a long-standing national effort to track bird population changes along hundreds of twenty-five-mile-long routes using standardized techniques. The routes are fixed in location, and most are outside of the city, which likely accounts for the discrepancy with our observations.)

Because the adapters we studied rely on the removal of dense forest cover to produce open lands for grasses, shrubs, and the rich mosaic of resources found along forest edges, they also decline in abundance as such "early successional" habitats mature and revert to forest. Any early successional habitat that was on a Breeding Bird Survey route in the 1960s would gradually lose its appeal to our adapters as it became more of a forest and less of an opening or edge. Only with new disturbance—a fire, severe wind, flood, or volcanic eruption—would forest be converted back to habitat adapters favor. The odds are simply stacked against seeing an increase in adapters at any given location—maturation is certain, but disturbance is not. So by returning to the same location each year to count birds, as is done to estimate regional trends, it is almost inevitable that adapter numbers will go down, as will the proportion of routes that cover recently disturbed ground. This does not necessarily mean adapters are declining in other places; where there is disturbance, they should flourish.

A better assessment of population trends for adapters would come from standardized counts during the summer months in many cities and other naturally disturbed areas. The only such count is the annual tracking done by Project FeederWatch. In this effort, citizens count the birds they see in their yards and report the results to scientists at Cornell University's Laboratory of Ornithology. Tens of thousands of "citizen scientists" participate in this effort each year. Although some of our adapters are not evaluated through Project FeederWatch, most of those that are have increasing or stable population trends. Three species—pine siskin, rufous hummingbird, and song sparrow—are declining at feeders as well as along Breeding Bird Survey routes.

As the population of some birds increases and that of others decreases during the construction of a neighborhood, the relative makeup of the entire bird community changes. This was evident in the Virginia woods studied by John Aldrich, and it was striking to us as we counted birds around Seattle. Pacific wrens rule unsettled forests. In our reserves 14 percent of all birds were these diminutive wrens. Nearly one of every two birds in forests was either a Pacific wren, chestnut-backed chickadee, Pacific-slope flycatcher, Swainson's thrush, or American robin. During the early years of construction a new group takes control. Violet-green swallows are the kings of this subirdia, accounting for 8 percent of all birds living in developments shortly after the start of construction. One of every three birds is a swallow, American robin, spotted towhee, American crow, or chestnut-backed chickadee. As neighborhoods age, swallows and robins remain atop the list of most abundant suburban birds. Gradually, towhees and chickadees are bumped out of the top five by starlings and juncos, such that in established subdivisions one of every three birds is a swallow, robin, starling, crow, or junco.

It is a common feature of plant and animal communities, especially those in temperate climates, to be composed of a few very successful species and many very rare ones. This was evident to various degrees in each community we sampled. As forests are settled, rare species accumulate substantially faster than do successful dominants. Some of the rare exploiters and adapters are just getting a foothold in new, suitable habitats. Other avoiders are barely hanging on in the pieces of their former haunts that remain. Despite having many rare species, common suburban birds share the wealth; the most common species included only about one of every ten birds we encountered. In deeply urban places, such as the central business districts of large cities, this is not the case. There, nearly every bird is a pigeon, house sparrow, or starling. Not so in subirdia.

The changing nature of bird communities in our neighborhoods is a quick reaction to the dramatic changes settlement brings to the land. I'll never forget driving past a newly cleared lot just a few blocks from my home. I had surveyed the birds on that lot four times each summer for the previous twelve years. Ten years before, the trees were cleared and along with them went the Pacific wrens, warblers, and thrushes. Now an acre of lawn replaced the dense growth of lovely salmonberry. Evicted was a rich community of sparrows and towhees. For the next two years I drove past the lawn daily, and I only twice saw a bird on it. Always it was a starling. The lawn seemed to me only slightly greener than concrete, and even pets and children dared not tread upon its manicured surface.

Gaining perspective on change in subirdia requires an artist's eye and a scientist's acumen. Fortunately, Jack DeLap possesses both. As a Ph.D. student, Jack carefully analyzes annual satellite images to distill changes in the vegetation associated with urbanization. His artistic talent supports this work as he uses an electronic pen to highlight and code each pixel of lawn, concrete, and forest on the many images. (Jack's original drawings also appear throughout this book.) These efforts put cold statistics to the changes that I took personally.

Jack discovered that the creation of Seattle's neighborhoods converted forests to mosaics of built, open, and forested lands. The process of development cropped a landscape that was 80 percent forest into one that was half forest, a quarter grass, and about 20 percent buildings and roads. Where forests once held two water features, now subdivisions supported seven small ponds. The mixture of land was stunning. Before the start of our study, Jack measured 384 miles in each subdivision where forest abutted different types of land (meadows, existing built areas, and the like), but after development, this "edge" rose by 60 percent to a whopping 650 miles. In our study, most of the gains and losses of individual bird species could be linked back to this changing tapestry of subirdia. In some cases one species' success quickly becomes another species' nightmare.

Laura Farwell cranked up the volume as the previously recorded Pacific wren call boomed from a speaker she concealed in the nearby bushes. Like a heat-seeking missile, a perturbed Bewick's wren streaked in looking for a fight. That's right, a *Bewick's* wren responded to Laura's "playback experiment" although she intended to coax a territorial response from the resident Pacific wren. Laura was learning how hard it is to be a Pacific wren—the icon of un-settled northwestern forests—when your forest becomes a suburb. Urbaniza-tion not only reduces and isolates what forest remains, but also invites a new bully onto the block, the streetwise Bewick's wren.

Bewick's wrens thrive in the shaggier parts of cities and towns across the United States. I always enjoy watching them flag their long, cocked tails as they bounce among the undergrowth. They are big for northern wrens—a quarter again the size of a tiny Pacific wren. Their song, *t-t-riing-dling-ling-ling*, zips through the gray gloom of an urban spring day. Out west, Bewick's wrens are classic adapters, taking to new suburbs as soon as the trees begin to fall. They are never particularly abundant, but what they lack in numbers, they make up in chutzpah. Not only are they aggressive in the face of a challenge—both from within their own and from closely related species—as Laura discovered, but they are also known to puncture the eggs of a rival. Their occurrence in newly created subdivisions puts them in unusually close contact with their forest-dwelling brethren, and this proximity is a second punch from urbanization to the gut of the Pacific wren.

When we watch wildlife respond to urbanization, in both positive and negative ways, we tend to focus on the fortunes and failures of a single species. But as the birds of a forest adjust to human settlement, the entire community is reshuffled. The changing abundance of one species may ripple through the full web of life. The web is refashioned, as new strings are added and old ones are diminished or reconnected into a new architecture. For example, research-ers from Perth, Australia, noted that some of the decline in species diversity at Kings Park may have been due to the predatory prowess of a particular adapter,

44

the Australian raven. The Bewick's wren might be acting similarly in the suburbs around Seattle, so Laura dug in deeper and unraveled the interaction between Bewick's and Pacific wrens for her graduate thesis.

She quickly confirmed that as the thick native understory that is so important to Pacific wrens for nesting, feeding, and shelter is removed, their numbers plummet. In our developing subdivisions Pacific wren abundance is halved during the first five years of construction. But rather than leveling off after forest clearing occurred, Pacific wren numbers continued their downward slide—another 40 percent were lost during the next five years. Throughout this decline, Bewick's wrens increased threefold, and it was their aggressive actions that contributed to the free fall of Pacific wrens and especially to their virtual absence from older neighborhoods. Although both typically use different sorts of habitat—Pacific wrens seek cool, dense evergreen forests and Bewick's wrens prefer shrubby clearings—in new subdivisions Bewick's wrens creep into the forest and take up residence in typical Pacific wren habitat. As Bewick's wrens defend their new digs, they evict resident Pacific wrens. Pacific wrens are forced from territories where they likely raised young the previous year and from which their ancestors raised wrens for thousands of years. Most of the evictions are via vocal summons, but occasionally feathers fly as the larger Bewick's wrens give chase to the mousy Pacific wrens.

Suddenly my childhood question seemed more urgent. Not only were animals such as Pacific wrens forced out of their homes by the clearing that followed survey stakes, but also those that remained may still be pushed along by other species more adept at surviving in our gardens and parks. I was left wondering where the wrens went. Did they die, or did they move? Kara Whitaker left the marshes of her native Wisconsin to help answer the question as part of her doctoral research.

Pacific wrens are about the size of a thumb, the color of dirt, and rarely venture more than about three feet above the earth's surface. How in the heck was Kara going to track them through the temperate rainforest? Pacific wrens were too small to carry even the lightest radio tag. So, we banded (ringed) nearly two hundred of them and crashed through the brush, among the bulldozers, and between the new rows of houses hoping to find our marked needles in the proverbial haystack. This sort of mark-resighting study is common among ornithologists, but we'd be the first to use it to track the movements of birds in response to human settlement.

To catch each wren, we lured them into fine-meshed "mist" nets that we placed strategically throughout their territories. We quickly removed the entangled wrens and kept them calm in soft, dark bags as we prepared to affix to each of their tiny legs a unique combination of colored plastic rings. A uniquely numbered aluminum band issued by the federal Bird Banding Laboratory would accompany these color bands. The combination of bands gave each bird a personal identity that we could see with binoculars, thereby allowing us to note its place of residence without having to catch it again. It takes only a few minutes to tag each wren, and then the hard work begins.

Just as John Aldrich mapped the territories of forest birds in Virginia, we now mapped every location where we encountered each wren—banded or not—in each of our developing subdivisions, forest reserves, and established neighborhoods. Over the next twelve years we mapped more than seven hundred territories and determined how their placement varied from year to year. Though we tagged many wrens and studied many more territories, we resighted barely half of the breeders in reserves and only a third of them in developing subdivisions. Annual losses were high, in some instances because of death and in others because of the movement that so intrigued us.

An annual mortality rate of 50 percent is not unusual for small birds such as wrens, so the birds we failed to find in forest reserves likely perished from natural causes. The greater disappearance rate in developing suburbs suggested heightened mortality, greater movement, or some combination of both. Kara gave us good evidence that at least some wrens flee development and settle on distant territories where they continue to breed. All the wrens we resighted stayed within the square kilometer area (not quite half a square mile) where we first encountered them before construction. In five cases where we could precisely map their territories, we found that wrens moved an average in excess of the length of two football fields (720 feet) between years. To us that isn't far, but to a two-inch wren it's a haul. We saw short movements in direct response to partial clearing—the removal of a few cedar trees caused one wren to shift about a football field away. And we saw long moves in response to large clearings—an established male wren moved eight times as far to find a secluded forest when the earthmovers leveled the land abutting his territory. So my dad was at least partially correct. Some animals do move and find new homes in the pieces of paradise that remain around our developments.

We never saw the wrens that owned territories in forest reserves or established neighborhoods move like those in construction zones. In both places, Pacific wrens are real homebodies. Our precise comparisons of sixteen mapped territories suggested that forest and neighborhood wrens move only half as far, about one football field's length (250–340 feet), as do wrens in active developments.

Moving may be risky, difficult, and even impossible for some individuals, but life in the suburbs for forest-dependent species is challenging, and for males, at least, it is also lonely. The wrens whose territories were somewhat protected from clearing as developments were created stayed initially but often disappeared the following year. Several were unable to attract a mate to their modified land.

Movements of another avoider, the Swainson's thrush, were similar to those of the Pacific wren. As with wrens, thrushes moved their territories substantially more in areas with active construction than in less drastically disturbed forest reserves or established neighborhoods. They were, however, less mobile overall than the wrens. Thrushes moved only fifty meters (about 160 feet) between years in suburbs and reserves, and just over twice that in developing neighborhoods. Salmonberry, which quickly covers recently cleared land, was a favorite of the thrush and likely allowed experienced pairs to remain within their territories even as forests, which are essential to wrens, were felled.

In contrast to thrushes and wrens, two adapters held fast to their territories regardless of construction activity. Breeding song sparrows and spotted towhees rarely moved more than half a football field's length (about 150 feet) from year to year. And this tenacity to the home turf occurred in reserves, established neighborhoods, and neighborhoods under active construction.

Subirdia is the place many of us call home or work. Physically, it is a richly interwoven mixture of residential, commercial, and wilder land. Houses, allotments and gardens, derelict and vacant land, golf courses and other outdoor sports sites, cemeteries, schoolyards, highway and railway verges, municipal utility stations, business parks, and shopping centers occur among places dominated by natural vegetation such as greenways, river and stream corridors, parks and nature reserves, pipelines and powerlines, steep slopes, and quarries. In a variety of locales, natural vegetation constitutes one-third to two-thirds of subirdia. Functionally, subirdia is the confluence between city and country that promotes a mutual exchange of plants and animals. It is also a place where people from urban and rural cultures come together as neighbors, friends, and acquaintances. In so doing, we learn how varied is the human perception of nature.

In subirdia, we form emotional and intellectual connections with nature. This is a place where we can interpret our personal experiences and observations of the natural world. It is where biophilia, a love of all life, can grow. The adventure sites we encounter here give our children places for important free and uncontrolled activities. Enjoying nature in subirdia allows them to unplug from technology a bit and to improve their powers of focused and sustained concentration.

Our creative hand fashions a rich birdlife in subirdia, but it is not without struggle and loss. In my neighborhood, Pacific wrens are a casualty of urbanization, whereas Bewick's wrens, violet-green swallows, and dark-eyed juncos are among the conquerors. Around the world the story is repeated, with different actors taking on the same roles. Subirdia is an amalgamation of adapters, exploiters, and avoiders. Stewardship of these riches calls on our ability to provide the varied resources that attract adapters and maintain avoiders. Exploiters are guaranteed and need little special attention. As our modification of the land combines birds into new communities, we create interactions that have never before been seen and rekindle others that played out long ago. The interactive strands that link subirdia's birds into an ecological web are as varied as the myriad animals that reside there. Some strands in the web are deadly, but others are supportive. As we now more fully consider a variety of ecological interactions, we learn about each bird's place in the web of life and begin to see ourselves within that web as well.

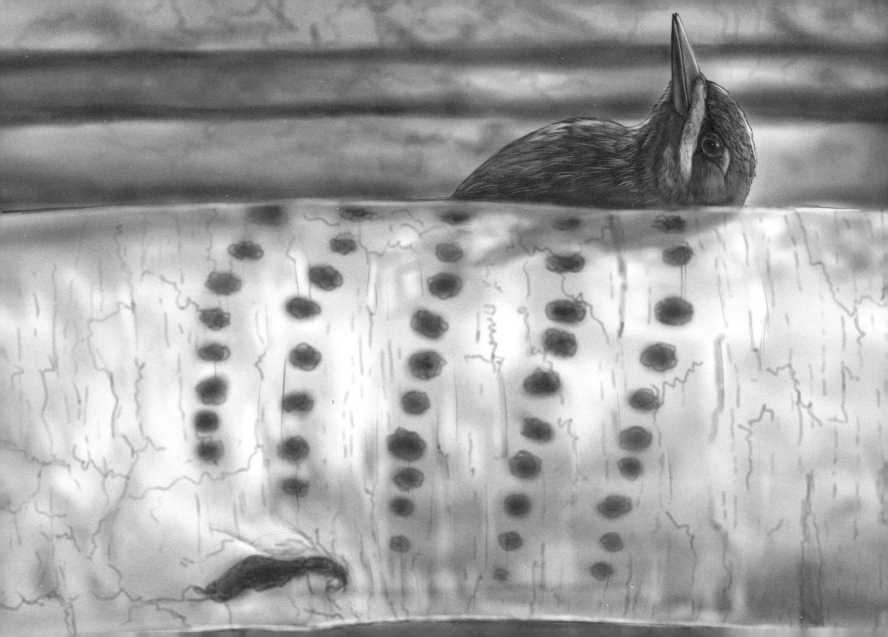

FOUR

A Shared Web

This feeling of connection to nature enlarges our understanding of community. . . . We emerge as stewards of nature motivated by an expanded appreciation of our personal and collective self-interest.

—Stephen R. Kellert, *Birthright* (2012)

Though relatively tame and constructed mostly from human hands, subirdia is not a zoo. There are no confining bars to isolate species or moats to separate humans from wild beasts. In subirdia, birds and humans share an ecosystem in which they are connected by fibers that define the web of life. This is a complex web that includes competitive battles among closely related wrens (such as the Pacific and Bewick's wrens in Chapter 3), and much more. Most of the songbirds we enjoy live in the web's center feeding on insects and plants from the web's wide base, while supporting a diverse range of predators that form the web's exterior. Connections that link organisms to the many plant and animal foods they consume define one aspect of the web's form, its "trophic structure." As these food chains layer predators atop midlevel consumers such as songbirds, the diversity and redundancy of species—the alternate

Facing page: *Red-breasted sapsucker with sap wells*

links in the chain—add stability. Species that broaden opportunities for others to join the web are important, as a keystone is important to the support of an arch. We humans are such keystone species when we diversify vegetation and supplement natural food, water, and shelter for birds. Some birds, "facilitators" in ecological parlance, also enable the presence of others, in theory by limiting dominant competitors and in practice by providing novel nesting sites. Come—let's explore the web!

Much of the ecological web exists out of sight—underground and in rotting wood. There, molds, bacteria, fungi, and a world of invertebrates convert the last molecules of sun-derived plant sugar to new life. These organisms are technically "decomposers," but functionally they are among the greatest of creators. Their bodies and chemical waste products provide us with an essential ecological service: soil, the foundation of terrestrial life. Some decomposers, such as termites, sustain subirdia in a more direct way.

Rust-colored, winged kings and queens wafted high on a warm August evening looking for sex and a place to settle down. Hoping that the orgy of inch-long termites would ignore my house of wood, I watched as they took their nuptial flight from my forest. All summer their colonies were turning my deadwood into rich humus. The munching of their siblings attracted woodpeckers. This evening members of the elite reproductive class that had avoided the long, sticky tongues of pileated, hairy, and downy woodpeckers were providing the foundation to an even wider part of subirdia's ecological web. Patrols of violet-green swallows ploughed through the clouds of weak flying termites like whales through krill. A lone Vaux's swift raced along on rapid wingbeats as it gorged on the insects. Less agile cedar waxwings hawked their share by darting from the tops of fir trees and hovering among the termite masses, beaking one bug at a time. The birds are not alone in this feast. A bat, most likely one of the many *Myotis* species that live here, assisted by its highly evolved sonar, had no trouble filling up fast on the fatty isopterans. I also saw green darner dragonflies, which remind me of miniature army helicopters, cut rectangular circuits among the termites, plucking them from midair with strong legs and crushing them with powerful jaws. No wonder this darner is the Washington state insect!

Keeping adequate stores of deadwood, given the foundational role it plays in ecosystems, is a challenge in neighborhoods. Not everyone enjoys an evening termite flight, and it is not uncommon for trees in my neighborhood to blow down in strong storms, damaging houses and cars. A few years ago my driveway was crowded with the remains of two great snags felled by a neighbor out of fear or for aesthetic reasons. I had often watched flickers, hairy woodpeckers, and pileated woodpeckers work these very trunks. Now they would fuel my fire rather than enrich the soil and make termites. They deserved the blow of a woodpecker's bill, not that of my maul. The actions of my neighbor are typical and in some situations justified, but our intolerance of dead vegetation severely limits one of the most creative forces in subirdia.

Woodpeckers are natural engineers whose abandoned nest and roost cavities facilitate a great diversity of life, including birds, mammals, invertebrates, and many fungi, moss, and lichens. Without woodpeckers, birds such as chickadees and tits, swallows and martins, bluebirds, some flycatchers, nuthatches, wood ducks, hooded mergansers, and small owls (screech, sawwhet, and pygmy) would be homeless. Some of these "secondary cavity nesters" are able to excavate nests and roosts in very rotten wood, and others might find a natural cavity, but homes in punky wood are easy pickings for strong-armed raccoons and opossums. Some also use nest boxes that we provide. But regardless of what we do for hole users, hole creators still need dead trees. In Seattle these critical ecological resources, while present in forest remnants, are nearly absent from built areas, severely limiting the abundance of woodpeckers. Even in suburban forests where small snags are plentiful, large ones such as those preferred by pileated woodpeckers are rare—fewer than two per acre remain.

Woodpeckers are never overly abundant; their requirement for a continuous supply of standing and downed deadwood is met in only large territories. Across their spacious ranges, however, they produce many cavities during their lifetimes—new nest cavities each year and numerous roosts. Although woodpeckers are intolerant of members of their own species, the defensive drumming and calling does not repel other types. Outside Seattle, the snags of a suburban forest can simultaneously support a pair of large pileated, medium-sized hairy, and small downy woodpeckers. Flickers, nearly the size of pileated woodpeckers, nest on the forest edges and feed on the ants and ground beetles often found in grassy lawns. Sapsuckers, midway between a hairy and a downy woodpecker in size, also coexist in this diverse cavity-producing guild. They avoid competition with the other woodpeckers by digging nests and roosts in live trees and drilling sap wells from which they feed.

The variously sized woodpeckers provide an array of cavities that differ in accessibility to and suitability for other species. Flicker holes are large enough for bluebirds but too roomy for chickadees. Pileated cavities support red squirrels, owls, and ducks. Tiny downy woodpecker dormitories are suitable for chickadees and nuthatches. Sapsuckers provide cavities for swallows. In this way woodpecker diversity magnifies overall bird and mammal diversity. Even the sap wells—horizontal rows of shallow taps that sapsuckers bore into live fruit, maple, birch, and pine trees—facilitate the biological diversity of subirdia by providing syrupy food for a host of insects, mammals, and other birds, in addition to the sapsucker.

Large species are often the first ones to be extinguished from the places we inhabit. Living large frequently bumps them up against our activities, making it difficult for them to meet their needs. So the sight in my backyard of a crow-sized woodpecker—dazzling black and white with a flamboyant, red head crest—made me pause and wonder how this could be. How could pileated woodpeckers survive here? This species was often studied in remote forests where its ecological importance as a facilitator was rightly recognized as sensitive to timber harvest. The other large American woodpecker, the ivory-billed, had already succumbed to our timber greed, and some scientists expected the pileated to follow suit. Yet here it was, digging carpenter ants from a dead stump, knocking back red huckleberries and ink-black cascara berries, and mooching suet from my feeder. And this was typical; pileateds live in just about all forested suburbs of the United States, from New England to Florida and westward to the Pacific. To learn more about the largest extant woodpecker in North America required some hands-on work and a willing graduate student.

The woodpecker decoy was a poor representation of the real bird. It was short and squat, but the color was true, and the red feathers scavenged from my fly-tying supplies gave some life to the Styrofoam bird, especially in a breeze. I watched from a distance as Jorge Tomasevic, an eager student willing to live among the woodpeckers, cued the audio player. Territorial and contact calls rang forth: *Peee, peee, peee, peee, peee, peee, peee! Wa, wa, wa, wa, waak, waak, waak, waak, waak! Chachuck, chachuck, chachuck.* Then drumming: an increasingly rapid *bop, bop, bop, bop, bop, brrrrrrrt* as the recorded bird rapped a resonant branch with its strong bill. The broadcast awoke the sleepy forest. Minutes later, Jorge was running to the net we had strung by the decoy to gather up his first subject: a male none too happy with the decoy, the recording, or the human. By the time I got to them, the powerful bird had bloodied Jorge's hands, but it mattered not. The Fulbright Fellow from Chile was beaming like a kid on Christmas morning. His hard work provided a great gift. Now, if we could just keep the bird from pulverizing our hands and quiet its ear-piercing calls long enough to attach a radio transmitter, we'd be all set.

Jorge calmed the bird with a soft hood as it gripped his coat like it would a mossy branch. We worked quickly, slipping its stiff wings into the harness that held the transmitter. Adjusting the fit of the harness's shoulder straps to be snug but not too constraining assured us that the bird could safely navigate its wooded world while wearing its new backpack. Though weighing less than 3 percent of the bird's mass, the transmitter would function for more than a year, giving us an accurate look at the bird's survival, reproduction, territory, and resource needs. This information would reveal how a potentially vulnerable bird managed to live in subirdia and what homeowners and resource managers could do to ensure its continued ability to do so.

Every day brought new insights into the pileated's world. Unlike the wild forests where the species was previously studied, in subirdia the woodpeckers flew across interstate highways and through expansive developments. Human activity on the landscape did not stifle them. In fact, they spent considerable time at bird feeders, sometimes bivouacking for the night outside a cavity in the deep bark furrows of big-leaf maple trees, and even breeding in cavities they drilled in snags that homeowners created by topping large trees in their backyards. Most nests and roosts were in forest remnants, but the overall size of a suburban territory was barely half as large as a wildland territory. To Jorge this suggested that subirdia was indeed rich in the foods and settings that woodpeckers needed. After five years of tracking the movements of two dozen woodpeckers in and out of suburban settings, Jorge demonstrated that survival and reproduction equaled that of the woodpeckers in wild settings far from human action. These magnificent creatures took advantage of suburban subsidies while clinging on to the vestiges of a wild and forested past. They adapted their diet and movements to the new humanized landscape in a way that the ivory-billed could not. In doing so, they lured many residents into their ecological web by inspiring customized suet production, tree maintenance, and school projects that linked students with charismatic nature.

In subirdia humans augment the actions of woodpeckers by offering nest boxes that facilitate secondary cavity nesters. Providing nest boxes is a common practice, especially in Europe. Barbara Clucas and Sonja Kübler, freshly minted Ph.D.s from the United States and Germany, respectively, studied this strand in the urban ecological web in Berlin, Germany, and Seattle. Teaming up, they traded binoculars for interview scripts and started knocking on doors. After nearly six hundred discussions, some with more than interesting residents, Barbara and Sonja discovered that one in three Berliners and one in five Seattleites had a birdhouse. Nest boxes were most common in suburban and rural neighborhoods and less frequent in the core of the cities. Provisioning birds with nest boxes forged another link between residents and the birds that shared their ecological web. Neighborhoods with strong connections—where people frequently said they provided boxes—had substantially greater numbers and a greater variety of secondary cavity nesters than did neighborhoods with weak connections.

In the heart of the city nest boxes and the inadvertent cavities provided in our built structures support native and nonnative species. Starlings and house sparrows are frequent beneficiaries of these cavities—both in their native Europe and elsewhere, including the more urban parts of Seattle. In suburbs, nest boxes are readily and successfully used by native species. Chickadees and Bewick's wrens flock to them. The ability of violet-green swallows to dominate subdivisions is in large part due to their use of human-made nest cavities in boxes, soffits, and streetlamps.

The use of nest boxes and built structures by nonnative species such as starlings could preclude nesting by native species. Starlings aggressively defend their nest cavities from a wide range of species. Sometimes they even evict other cavity nesters from their own holes. In Mississippi, native red-bellied woodpeckers lost more than half of their nests to starlings. Flickers are also common victims of starling abuse. Usurped flickers often renest, but these later attempts produce smaller broods than the lost first attempts. As a result, flickers' productivity suffers near starlings.

Although starlings have the potential to limit populations of many cavity nesters, in practice this is rare. Flickers and red-bellied woodpeckers produce cavities of just the right size for starlings, so they are frequent and conspicuous targets for takeovers. While some nests of these species fail because of starlings, over their entire range no ill effect of starling invasions is yet apparent. In fact, of twenty-seven native cavity-nesting birds in the United States that live among starlings, only sapsuckers show a rangewide decline in abundance that researchers attribute to starling aggression. Jorge's research in Seattle suggests why the detrimental effects of starlings are limited; most live deep in the city, and nearly every one of the 120 starling nests we observed was in a utility pole or building. Nesting in our homes is an old tradition of starlings—a century ago W. H. Hudson found them nesting under a zinc strip covering a crack in the wall of his English village home—and we suspect that it may well keep them from directly competing with many of our native species.

Species that facilitate the presence of others, such as woodpeckers do by creating cavities in trees, may be especially important in moderately disturbed environments, such as a city's suburbs. Accordingly, subirdia's riches may be particularly reliant on positive interactions among species that maintain and extend diversity. As we remove old trees, trim dead branches, and generally limit the natural decay process that chickadees, nuthatches, and creepers require, we increase the importance of natural and novel actions to facilitate the continued existence of these birds. Fostering woodpeckers by leaving and recruiting snags in forest remnants naturally supports secondary cavity nesters and keeps subirdia diverse. Placing nest boxes in our yards where snags are much less common extends this support from the forest to a great diversity of native and nonnative adapters and exploiters of subirdia. One of the intimate connections between humans and the suburban web of life is also revealed. We, like woodpeckers, can be facilitators of diversity.

Considering ourselves as facilitators, not simply destroyers, of biological diversity shines a new light on our place in the web of life. Some of our actions enable other species to thrive where they would otherwise be rare. It is this role that supports the diversity I often encounter around people's homes, and allows it to grow. When our facilitative relationship with nature also benefits us, then it may flourish and be sustained. Those of us who retain dead trees or place nest boxes in our yards enjoy the wonder of watching woodpeckers listen and dig for termites; we are serenaded by wrens; and we benefit from the appetites of swallow, chickadee, bluebird, and flycatcher broods that are sated on insects, including pesky mosquitoes. Providing food and water for birds (see Chapter 5) is another facilitative link between people and birds that enhances diversity. Through facilitation, we become part of a natural mutualism benefiting other species that return the favor.

I take comfort in discovering ecological relationships that are mutual. What's not to like when everyone wins? However, like most field biologists I also love a good chase, a tough fight, and a front row seat to a killing.

An unusually bright December afternoon brought a small flock of pine siskins to my bird feeder. With my wife and daughter, we took a moment to enjoy their subtle yellow color and their aggressive posturing as they pushed each other aside for sunflower seeds. From a sudden gray and brown streak materialized a sharp-shinned hawk that also had been watching the feeder. Instantly, the hawk snared

Pine siskin caught by a sharp-shinned hawk

a siskin and fluttered to the ground, where its sharp talons and strong feet turned the songbird into a meal. We were saddened and simultaneously excited by the drama we had just seen. Nature can be cruel here in subirdia, as everywhere. But to an ecologist, predation is a sign of health that indicates an ecosystem with ample prey to sustain and lengthen the food chains built atop them. Death is not easy to watch, but it is a fundamental feature of ecological webs.

One might think that predator-prey links are rare in the suburban ecological web, but that would be wrong. True, humans have little tolerance for large predators near their homes; grizzlies, wolves, tigers, lions, and their ilk are often excluded. But smaller predators often thrive among us, in part because we remove their enemies and in part because of the abundance of prey that inhabits subirdia. In my neighborhood red-tailed hawks and raccoons fatten up on baby crows whose parents have likely robbed eggs from the nest of nearby Steller's jays that regularly eat the eggs and young of other songbirds. Garter snakes prey on the eggs and chicks of ground-nesting birds such as the dark-eyed junco and Wilson's warbler. Coyotes and bobcats mostly stay hidden, but they are common.

One of my most memorable sights occurred as I was searching for nests early one morning. The neighborhood I was in featured many wooded trails, and as I walked down one I heard a commotion of crows coming my way. Their mobbing calls told me I was about to intersect a predator, so I sat and waited. As the calls grew louder, I could see the birds diving toward the ground. Suddenly a regal bobcat, stubby tail held high, pranced by. Its entourage of crows trailed behind, a bad omen to any unsuspecting rodent or ground-nesting bird, but a thrill to me.

Raptors—the term we apply to hawks and owls—seem especially able to live in cities. Peregrine falcons have nested in cities and towns since the Middle Ages, feasting on pigeons and other moderate-sized birds. Their populations plummeted in the mid-twentieth century because of pesticides such as DDT. Banning DDT and dedicated restoration efforts, however, have brought the falcons back to more than sixty urban centers in the United States alone. They eat mostly pigeons, but also much more—juncos, jays, woodcocks, and whip-poor-wills. They are not dainty eaters, as I learned at a University of Washington football game. Above my seat, high in the steel girders of the stadium roof, perched a peregrine. Without flying, it scampered along the I-beams and deftly snatched a pigeon. As it plucked the bird, feathers floated down like dry snowflakes onto the fans engrossed in the game. After an hour the full bird roused its plumage and began bobbing its head in preparation for a flight. A sated bird always lightens its load before flying, and this peregrine was no exception. It cocked its tail and sent a missile of whitewash onto the crowd. Fortunately, my seat was just out of range.

Peregrines aren't the only raptors in urban areas. Bald eagles, burrowing owls, eastern screech-owls, osprey, and Mississippi kites reproduce at high rates

Red-tailed hawk mobbed by crows

in suburbs. European goshawks are fully urbanized. Spanish imperial eagles and Swainson's hawks, in contrast, survive in some cities but reproduce poorly.

Some raptors adjust to urban environments by expanding their diets. In western Europe, tawny owls and kestrels that live in cities switch from a typical diet of mammals to one rich in birds, especially house sparrows. Kestrels also scavenge human refuse, cleaning the meat off of steak bones and snacking on sandwiches left behind in schoolyards.

Stan Rullman knew he wanted to be a biologist when he was seven years old. Raptors were a special draw to the young lad's imagination, and he spent many a childhood day searching for their nests. Thirty years later, despite having studied all manner of life as a curator at the Cincinnati Zoo, Stan was ready to pursue an advanced academic degree. We quickly decided that the hawks and owls around Seattle would be perfect subjects. To understand diurnal and nocturnal raptors was challenging. Stan would frequently stay up all night blaring owl calls in a neighborhood and then walk the streets the following day playing the screams of hawks. Through his dedication we learned that most neighborhoods harbored three to five species of raptors. Stan found that one, the barn owl, was an exploiter. This white, monkey-faced species was found only in the most urban areas. Western screech-owls also were closely associated with development, never responding to calls in subdivisions that were mostly forested. The pygmy owl, on the other hand, was an avoider. This little guy was heard only in neighborhoods and reserves dominated by forest. To our surprise, most raptors were adapters that lived in nearly every neighborhood. These included Cooper's, red-tailed, and sharp-shinned hawks as well as barred and great horned owls. The Cooper's hawk seemed special.

"Coops" are common in suburbs across North America. They belong to the genus *Accipiter*, which includes the larger goshawk and the smaller sharp-shinned hawk that raided my feeder. The short, rounded wings and long, dexterous tails of all *Accipiters* allow them to outmaneuver most of what they chase. Their hunting style is to sit, wait, and surprise their quarry, often at feeding, watering, and roosting sites. They are opportunistic—robbing nests, scavenging on the dead, and ambushing the unsuspecting. I guess Coops are a close second to the domestic cat in being a songbird's worse nightmare.

Seattle is stuffed with Coops. They hunt from my backyard tiki torch. I see them among the high-rise buildings of the city center as well as the lofty peaks of the high Cascades. They regularly build their large stick nests in the quieter parts of our neighborhoods and strafe nearby feeders. It takes several hundred songbirds to raise a brood of hungry hawks each spring, and Stan wondered what living next to a Coop nest might mean to the main items on the menu. It turns out that about a third of the places where we counted songbirds and monitored their reproduction and survival hosted Coop nests; the other two-thirds did not. Stan took advantage of this information to calculate the effects.

It was indeed dangerous for prey to breed near the nest of a Cooper's hawk. Stan estimated that the annual survival of robins and Swainson's thrushes—favorite Coop prey—was about 7 percent lower as a result of nesting near a hawk nest. Compounded over several years of a bird's expected life, this greater mortality could reduce a population. The breeding success of prey was even worse; it dropped from a fifty-fifty chance of fledging young in hawk-free places to a one-in-three chance of doing so where hawks nested. But this is where things got a bit more complex, and interesting.

Not surprisingly, nesting success of all songbirds increased with distance away from an active Coop nest. But this increase only lasted for a distance of a few city blocks. Beyond that, success again declined. This uptick in nest failure is probably indirectly related to the hawks' presence. Near their nests, hawks shield songbird chicks and eggs from predation by other predators such as jays, crows, squirrels, and chipmunks. These generalist predators are rare near hawk nests but increasingly common as one travels away from the nest. As the protective umbrella of the hawk drops away, crows and squirrels prey on robin and thrush nests, leading to an increased risk of nest failure far from hawk nests. Failure very near hawk nests is likely due to actual predation by the hawks on the nestling songbirds. As with most things in life, moderation with respect to nesting near a Cooper's hawk nest is the best strategy.

Despite their effects on the predators and prey that nest near them, Cooper's hawks didn't overly influence the organization of Seattle's suburban bird communities. Their presence was unrelated to overall diversity, though it did provide some check on the dominance of the robins and Swainson's thrushes.

When species form new communities, as birds do in response to the creation of cities and suburbs, they may find themselves in the company of ancestral foes. Through evolutionary time, similar species that require the same, rare resources compete. This interspecific competition is thought to eliminate inferior species and also foster divergence. Because of competition, similar species may come to differ in critical aspects of their structure, such as the size or shape of their beaks, as happened with the famous Galapagos finches and splendid Hawaiian honeycreepers. The habitats where they live and their behavior, such as the tactics they use to catch a meal, can also be shaped by competition. This "character displacement" may happen gradually or rapidly during brief and infrequent ecological crunches, for instance, during years when critical seed resources are especially rare. Because interspecific competition works to diversify close competitors over evolutionary time scales—typically thousands to millions of years—species that have evolved together in the past often show little present-day competition. In a sense, battles long ago have already settled the score. Species that coexist today can be thought of as differing because of past competitive interactions; the places where they live and nest, the foods they eat, and their morphology may all have been visited by what community ecologist Joseph Connell calls the ghost of competition past. When humans disrupt the barriers that evolution has put in place to reduce competition, the specters of competition present and future pay a visit and shock subirdia's ecological web.

Noisy miners are indigenous Australian birds with bright yellow beaks and black masks. They are about as big as a robin and somewhat resemble a myna bird with scalloped gray, black, and olive plumage. Like Bewick's wrens in Seattle, in the town of Crows Nest, Australia, noisy miners are bullies. In this part of Queensland, about one hundred miles west of Brisbane, agriculture and some urban development fracture and reduce native eucalypt woodlands and the dense understory of shrubs. This modification suits the miner well, and its abundance has increased greatly throughout eastern Australia. A host of other woodland-dependent songbirds whose numbers have declined are being revisited by the ghost of competition past, and it has a big yellow beak.

Noisy miners aggressively defend their territories from other species. Presumably through evolutionary time, species excluded by miners settled in habitats where these aggressors were rare, such as expanses of eucalypt woods where native shrubs were dense. Urbanization and agriculture have degraded the shrubs that many forest birds need in order to coexist with miners. Smaller birds move more easily through the shrubs than do the larger miners, finding food and nesting space ample within them. Miners forage in open ground and remain bit-part players in the bird community where shrubs predominate. But competition is back at work in Australian suburbs, pitting miners against birds such as the superb fairy-wren and silvereye. This time these small songbirds have few shrubs in which to hide. Competition may force some birds to evolve new distinctions that again remove them from the miner's tyranny.

In subirdia, new ecological interactions such as the renewed competition imposed on fairy-wrens by noisy miners are precursors to inevitable evolutionary change. Adaptation, an adjustment to new situations such as the presence of an aggressor, is a creative evolutionary response that gave us the wonderful diversity of birds that today coexist in shrubby eucalypt woodlands. Extinction is also an evolutionary response to new situations, and it is down this path that many Australian scientists believe the miner is taking woodland-dependent songbirds. Restoration of native shrubs in suburbs is proposed as a strategy to exorcise the ghosts of competition before they extinguish the fairy-wren and others.

Bewick's wrens and noisy miners tell me that subirdia's diversity is a work in progress, a melting pot of sorts that brings together animals from many formerly distinct walks of life. Some of their differences evolved in the past when their ancestors competed. Our actions are bringing some back into the competitive ring. By opening forests or planting trees in grasslands and deserts, we bring together in suburbs species from open lands and those more typically found in forests. Whenever new species come into competition, we can expect the larger or more aggressive species to exclude the meek and weak. Almost certainly this exclusion will reduce diversity. Setting aside refuges and restoring habitat features that allow competitors to coexist are simple ways to keep subirdia diverse. By keeping different suburbs distinct, we can also allow species with unique requirements to live somewhat isolated from each other. For example, in central Bohemia, Czech Republic, black redstarts prefer open, densely built parts of the city, while redstarts prefer places with more trees. Black redstarts dominate redstarts, but this is rare because of the mosaic nature of open and tree-covered habitats in the city. Keeping open and forested neighborhoods distinct and separate allows cities in Bohemia, such as Breznice, to host two, rather than a single, species of redstart.

When noisy miners, black redstarts, and Bewick's wrens displace other native species, I am concerned but take some consolation in the fact that these aggressive and successful species too are natives. Unfortunately, interactions with nonnative species pose a much more serious risk to biological diversity. Nonnative predators have decimated island bird communities around the globe. Nonnative competitors are common in plant and invertebrate communities where they are a major factor in recent extinctions. Such competitors seem to pose less of a problem to birds. Black drongos may challenge some birds of Micronesia; barred owls obliterate spotted owls. But in general, I think that the many parrots, mynas, upland game birds, and songbirds that have been introduced around the world coexist with natives as benign or even beneficial partners, especially in food-rich urban areas. An exotic dove, however, is changing my mind.

Eurasian collared doves exploit cities and towns throughout their native ranges in Asia and Europe. Since their invasion of Florida in the early 1980s, they have been well along the way to doing the same in the United States. This nonnative pigeon of moderate size is well distributed across the southern states and is moving north. A small number of them even live in Seattle's subdivisions and neighboring towns, as far as one can get in the continental United States from Florida. Recently, they've been spotted even in Alaska. Several native doves coexist in the southern United States, raising concerns about whether there is room and food enough for one more.

In Socorro, New Mexico, high in the Chihuahuan Desert, mourning doves were a common suburban bird twenty years ago. That was before larger white-winged doves expanded their ranges northward and began breeding in Socorro. With the arrival of the white-wings, mourning doves shifted their distribution from residential areas to the surrounding desert. Inca doves, which are small ground-feeding birds, also moved north and into Socorro's residential area at this time. In some places one could find all three doves coexisting, but by 2006 most residential areas held only Inca and white-winged doves. That is when Eurasian collared doves became increasingly common. White-winged doves dominate collared doves and have remained abundant in New Mexican subdivisions. But Inca doves have been unable to do so, declining in number shortly after the Eurasian invasion and disappearing from some areas five short years later.

In Florida a similar story is unfolding, though scientists are quick to point out a positive association between their native doves and the invaders. All doves—mourning, ground, white-winged, and collared—co-occur in residential areas where food is abundant—Florida's version of subirdia. But during the first decade of the twenty-first century, the numbers of Eurasian and white-winged doves, both invaders in Florida as in New Mexico, continued to increase while native mourning and ground doves (a small species closely related to the Inca dove) declined. That the native doves remain in Florida's suburbs may have everything to do with climate.

The high elevation of New Mexico imparts a strong seasonality to its climate. Snow and freezing temperatures are common in Socorro during the winter. These challenges to ground-feeding birds such as doves are rare in subtropical Florida. During these ecological crunches competition may be severe. Perhaps this is why New Mexican white-wings were able to oust mourning doves, and Inca doves have been excluded from the northern portions of their range. Although this is a reasonable scenario, we cannot be sure whether, or how important, competition has been in restructuring the dove community of Socorro. It is possible, for example, that cold winters alone doomed the Inca dove. In warmer cities of southern Arizona, Inca dove numbers are actually increasing. As temperatures drop, these small animals stay warm by huddling—actually forming pyramids that keep the central birds especially warm. Every five minutes or so the birds reshuffle; inner birds move out and outer birds move in so that the group can survive. (That's something I'd like to see!) Unfortunately, at even moderately low temperatures (less than twenty-one degrees Fahrenheit) Inca doves die. The true interactive effects of climate and invasive species on native doves may not be known for some time, but the complexity of the relationship highlights the dynamic nature of subirdia's diversity. Far from being static, the communities of birds that live in our neighborhoods are constantly adjusting to new neighbors that may facilitate or inhibit the presence of others.

Although interspecific competition may be the most elusive interaction to document in today's subirdia, its potential to affect the assembly is large. In

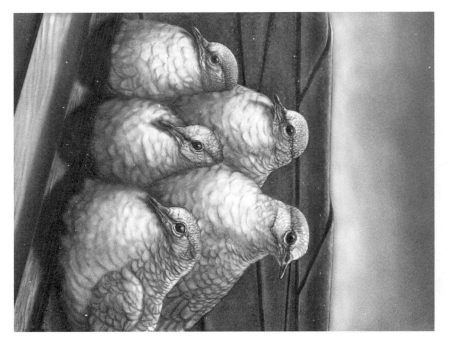

Pyramiding behavior of Inca doves

Seattle's suburbs we regularly document a community with five facilitators, three foreign invaders, seven predators, and dozens of potential competitors. I wonder how house finches will affect purple finches, how white-crowned sparrows will affect song sparrows, and how Anna's hummingbirds and rufous hummingbirds will get along. As long as our plantings continue to produce nectar-rich flowers and fruits, our bird feeders remain well stocked, and our city streets aren't too tidy, I suspect that abundant food and generally mild urban climates will keep competition's influence subdued.

The diversity that subirdia contains may seem redundant, but each piece helps maintain the whole. A diverse ecosystem performs better than does a simple one. Collectively, the many species in a wide ecological web capture more energy from the sun's nourishing rays and convert it into the vegetation that ultimately supports us all. A greater number of species is better able to recycle the plant-derived nutrients that move through the ecosystem, especially its decomposers. In variable and changing conditions, diversity has both a stabilizing and a performance-enhancing effect. In a diverse community, as some species decline in response to change, others increase. Challenges posed by aggressive nonnative invaders, many of which are devastating to human economies, are more likely stifled by the collective actions of a diverse rather than a simple ecosystem. Species diversity can be thought of as an insurance policy against today's environmental change and tomorrow's new evolutionary challenges.

For subirdia's insurance policy to pay its full dividend, diversity must not only be assembled, it must be sustained. The mere presence of a species is not always indicative of its staying power. To gauge the ability of birds to persist in subirdia, we must track the life trajectories of individuals. Their stories are powerful.

On a July day in 2009, a small, darkly streaked, plump song sparrow sings from a subdivision midway between the urban and wild ends of the Seattle

Song sparrow

urban gradient. Three brisk notes, a trill, and a punctual finish raise new questions. Each male song sparrow builds a repertoire from the utterings of his father and neighbors but develops his own characteristic voice. This one sounds familiar to me, and it is ringing out from a brushy spot where I have monitored other song sparrows. I stalk him to determine whether it is a bird I have banded and to see how he is adjusting to suburban life. It turns out that he is a bird I know well, with the blue and green plastic bands I placed on him four years ago. Each of the past three years he sang all spring and summer within this territory and failed to entice a female partner. His reproductive output, as best as I could discern, was zip. Yet maybe because he had no other option, he clung to his territory as the forest around him was converted to a new subdivision. He remained after a northern flicker and Pacific wren were forced out. People cut down the flicker's nest tree and cleaned up the protective brush that held the wren. Feeders may have helped sustain the sparrow, but he paid a price. I smile this year as I find he has a mate, and together they succeed in raising at least two fledglings.

Is it enough for suburban birds to occasionally win the reproductive lottery? To find out how successful they must be to leave a sustainable legacy requires us to measure their lifetime reproductive output. It's time to hit the bushes and look for nests.

The loon is a diver; the cormorant a fisher; the petrel a mariner . . . the bush-tits belong to the builder's caste. They are specialists in domestic architecture.

—W. Leon Dawson with John H. Bowles,
The Birds of Washington (1909)

By the first week of June 2002, I'd already passed the eighty-foot-tall Douglas-fir ten times. Flocks of bushtits roamed this area earlier in the year, and lately I'd seen a pair of them. I searched in vain for their nest. Bushtits, tiny gray cotton balls that float among the thick salmonberry, intrigued me. Like penguins amassing on the ice before plunging into dangerous waters, they gather at the slightest gap in cover and then singly bounce across, twittering in celebration as they reconnect on the other side. This day, I wondered about their social dynamics. Does the flock completely dissolve into pairs to breed? Why are the males' eyes beady black while the females' are shiny gold? In my distraction the pair I was stalking disappeared and with them any reasonable chance of scoring the nest.

Facing page: Bushtit

A sizeable fraction of ornithology involves watching nests. For me, in this neighborhood under construction, finding nests and following their fates was the first step in understanding whether the amazing diversity of subirdia was sustainable. I was dubious; regardless of where a bird nests, producing free-flying young from a set of eggs is rare. I'd often watched predators take a nest, heard from residents about the loss of their favorite robin brood, and learned about the frequent predation on urban nests from the writings of my colleagues. All this information suggested that any eggs lucky enough to be laid in an urban nest would be quickly slurped up by roving jays, rooks, currawongs, or crows. If an egg survived undetected long enough to hatch, then a marauding squirrel, coon, stoat, or fox would gladly add it to its general diet. In the rare event that a chick grew to the point that it was able to stretch its wings and fledge from the confines of the dangerous nest, then the neighborhood tabby would dispatch it and proudly trophy it home to its owner. To see all this drama, I had to find nests and monitor their output, so I looked in the bushes and on the ground, scanned the tree canopy with my binoculars, and waded deep into the briars to examine any clump that might conceal a cup of eggs.

Despite my inability to find the bushtit nest, it had been a good first month of the nest-searching season. By looking in likely places and attending to the behavior of parental birds—noting especially where they carried beaks full of nesting materials or wads of food for chicks—I'd already found sixteen nests of eight different species. I checked on each one every few days. When the parents were away, I'd peek in from a distance with an old car side-view mirror fastened to a stick to count eggs or chicks. Afraid of attracting undo attention to my prize, I never marked a nest site in the field, but I recorded each one's location on a paper map and also consigned it to memory. As expected by now, several had failed; the previous week a Douglas squirrel poked its head out of a tree cavity where a few days earlier I had listened to the begging cries of a red-breasted sapsucker's growing brood. Probably the same hungry squirrel also silenced nearby nests of a Steller's jay and a brown creeper. Both Pacific-slope flycatcher nests I'd found were abandoned after a single egg was punctured, perhaps by the teeth of a deer mouse or the beak of a parasitic cowbird. But many nests had also fledged. Pacific wren, chestnut-backed chickadee, American robin, and red-breasted nuthatch families were clamoring after their parents for food. Time was running short if I was to find the bushtits before they fledged.

It was the vibration that finally caught my eye. The tip of an outstretched fir branch seemed to droop a little more than I had remembered, and it was quivering! As I trained my binoculars on it, a male bushtit popped out. There, finally, was the nest; a tightly woven, pendulous stocking of moss that blended perfectly into the emerald tip of the bough. A few minutes later the golden-eyed female came in with a huge caterpillar. My presence deterred her not in the least. As she disappeared into the mossy orb, it burst to life, shaking the old fir branch and bringing a smile to my tired face. Five days later, fledgling bushtits spilled forth like corks from a case of champagne.

During twelve years of searching, my students and I managed to find only forty-three bushtit nests. As enthralling as each discovery was, they answered only a few of my questions. Bushtit parents were quite successful, fledging young from nearly six out of every ten nests. They were especially successful in established neighborhoods and least successful in forest reserves—just as expected for a bushtit's success, because the eggs and chicks developed in secret, deep within the gourdlike nests. Occasionally, we counted fledglings as they huddled shoulder to shoulder on a branch. But this appearance was a rare sight. To estimate the annual number of young produced by a nesting pair of birds required a more cooperative subject. Few ornithologists would argue that if one seeks nests, then a thrush—be it an American robin; European blackbird; fieldfare; or song, Swainson's, or wood thrush—is the ideal subject.

It is easy to find nests when we form a "search image" for them. Training brains to filter out the typical jumble of shrubbery, leaves, and shadows to focus on the out-of-place clump, errant tuft of moss, and shady tree cavity, my team found more than sixteen hundred nests of forty species. We found nests of ducks and hummingbirds; nests of warblers and woodpeckers; and nests on the ground, behind walls, and well into the canopy. But half of all the nests we found were located around eye level in native understory shrubs. Their architects were American robins and Swainson's thrushes, both of which build similar, open-cup nests. Swainson's thrushes mostly built their palm-sized cup in a shady crotch of salmonberry. Robins were less choosy, placing their larger domiciles on nearly any suitable surface, natural or manufactured, including the garage rafters right above my mother-in-law's clean car. Thrushes construct their nests to blend into the background, which in Seattle means siding them with fresh, emerald moss. The moss often trails below the nest, like a Sasquatch's ratty beard, to disguise the stark outline. The other construction materials belie their builder's identity. Robins plaster a grass lining together with mud, but Swainson's thrushes forego mud and line the nest with a delicate bed of translucent leaf skeletons.

Despite their similar nesting habits, in terms of productivity, robins and Swainson's thrushes are quite distinct. Although twice the heft of a female Swainson's thrush, a mother robin lays fewer eggs. A typical clutch of her bright blue eggs numbers three, whereas the paler Swainson's eggs are nearly always laid in sets of four. Half of all Swainson's nests we found fledged young, but only 40 percent of robins' nests did so. Squirrels, jays, or hawks preyed on most of those that failed. A few were destroyed by storms or abandoned by parents that nested too close to trails frequented by humans. Nests that successfully fledged young birds rarely contained dud eggs or starving young, something not uncommon in other urban locales, and therefore typically fledged one baby bird for every egg laid. Adding up these differences means that on a nest-by-nest basis, Swainson's thrushes produce nearly one more fledgling than do robins. And they do so especially where they were most common—in forested reserves.

This result is where most studies of nesting success stop, and while thrushes afford us a large sample of nests with visible output, counting just one batch of eggs or fledglings is far from the full story. Because robins are short-distance migrants, often wintering near where they nest, they can start breeding earlier and continue longer than can Swainson's thrushes, which winter thousands of miles away in Central and South America. In Seattle, robins begin nesting in late March—four to six weeks before Swainson's thrushes. And robins nest well into August when Swainson's thrushes are already busy fueling up to fly south, not to lay more eggs. Female robins are also phenomenal egg machines, nesting again rapidly after one nest fails and occasionally succeeding at fledging two or more full broods within a single year. Swainson's thrushes nest again only after early failures and in our experience have only once successfully raised two broods in a single season. Scoring our two thrushes' season-long production reveals that both species are equally successful. Each pair of birds typically produce two young per year.

The range in annual productivity of robins and Swainson's thrushes and its composite measures—number of eggs laid, success of hatching, number of fledglings, and frequency of breeding—were typical of the birds we were able to study in detail. Although we found far fewer nests of other species, some were especially amenable to observation outside the nest, something much trickier for thrushes. Rather than risk attracting predators or mistakenly trampling upon nests of Wilson's warblers, Pacific and Bewick's wrens, dark-eyed juncos, song sparrows, and spotted towhees, we watched for signs of breeding as we mapped out a pair's territory. We noted courtship and carrying of nesting material and nestling food, and we listened for the raspy begging of a hungry, young fledgling as it trailed its parents. This protocol allowed us to determine the annual productivity of 8,201 pairs of birds, mostly without sighting—or disturbing—their nests. Like Swainson's thrushes and robins, each pair of towhees or Pacific wrens fledged two young per year. Juncos, song sparrows, and Bewick's wrens were a bit more successful, typically fledging three to four young per year.

The collection of species that we could study in detail—what we call our "focal species"—differed substantially in annual productivity within and around suburdia. Swainson's thrushes and Pacific wrens were most productive in forest reserves. Juncos were especially productive in developments. Robins, song sparrows, Bewick's wrens, and spotted towhees were equally successful in reserves and in neighborhoods that were undergoing construction. The ability of towhees to capitalize on nesting near forest edges was also noted by researchers in nearby Portland, Oregon.

Woodpeckers and birds that nest in the holes they drill into trees broadened our perspective a bit more. Tina Blewett and Jorge Tomasevic studied more than three hundred nests of native primary and secondary cavity nesters. Primary cavity nesters are what we call woodpeckers. Secondary cavity nesters include native species, such as chickadees, which nest in old woodpecker burrows, and nonnative species, such as the European starling and house sparrow, which nest in crevices, cracks, and holes in our houses and other built structures. Compared with open-cup nesters, cavity nesters had extremely high nesting success. Their relatively large and very noisy broods grew up in the security of a solid-walled house and were only infrequently preyed on. Snakes cannot climb up slick, metal poles to eat young violet-green swallows in weatherproof and warm nests placed in streetlamp fixtures high above the road. Even the most persistent raccoon cannot reach a brood of starlings that grows in a cement utility pole. Jorge and Tina found that only one in five native primary and secondary cavity nests failed to fledge young. This result is substantially better than the one-in-two average nest failure rate experienced by open nesters. Nonnative exploiters were even better; fewer than one in twenty of the house sparrow and starling nests we watched failed.

The ability of adapters and exploiters to reproduce in homes, yards, and even the nooks and crannies of nature that remain during major construction activities appears important to their ability to take advantage of the warmth, landscaping, and abundant food that subirdia offers. The inability of some avoiders, such as the Swainson's thrush, to do likewise may confine them to larger reserves rather than outright suburban settings.

My colleagues across North America and Europe are discovering much the same. Although individual bird nests often sustain other creatures higher on the food chain, species able to adapt to and exploit our yards and parks such as cardinals, house wrens, catbirds, magpies, chaffinches, blue tits, and mockingbirds make up their losses over the course of the summer. Only three of the ten studies that report annual reproductive output found it to be lower in urban than nonurban landscapes. By breeding early and often, urban birds regularly overcome handicaps such as smaller clutches of eggs, weaker nestlings, and frequent encounters with predators, parasites, and contaminants that lower a single nest's production.

Watching thousands of birds reproduce among us also makes it obvious that breeding is only part of what it takes to sustain a population. Pacific wrens avoided subdivisions, but when present there, they actually reproduced quite well; only the consummate adapters—dark-eyed juncos, song sparrows, and Bewick's wrens—did better. Northern flickers, the most common woodpecker in subdivisions, were the least successful primary cavity nesters, fledging young from just over half of their nests—partially because of the flickers' interactions with nonnative European starlings (see Chapter 4).

To thrive in cities, birds must harvest the resources we humans provide and avoid the obstacles that surround them. For adapters and exploiters the path is rather straightforward: find the feeder and avoid the feline.

Black-headed grosbeak with sunflower seed

Finding the feeder is easier than we might think. Somewhere between one in five and one in three families in Europe, North America, and Australia feed birds. Feeders are packed into densely settled parts of cities, even though per capita feeding rates are greatest in the countryside. Thistle and sunflower seeds are common offerings, but where hummingbirds hang out, such as on the West Coast of the United States, more than half the people who feed birds also provide nectar. In the dry Southwest deserts, more than 90 percent of feeders offer true urban oases with plenty of water and food.

An estimate of the amount of food people provide to birds is staggering. A typical participant in Cornell University's FeederWatch program dishes out more than three hundred pounds of seed and twenty pounds of suet each winter. Across the United States, one-half-million to nearly one-and-a-quarter-million tons of seed enters the avian food chain through bird feeders. That is roughly the same as the amount of corn, wheat, and rice that the U.S. government donates annually to feed people in Africa. At an annual cost of $3.5 billion, this offering fuels human as well as avian economies. It is no wonder that 350 species of birds have been reported eating at feeders in the United States.

Feeders enable migrating species to live farther north and at the same time lengthen the breeding season and enhance overwinter survival for many resident species. Cardinals, tufted titmice, and Anna's hummingbirds are marching north because of feeders. Suburban Florida scrub-jays, an endangered species, use protein-rich peanuts to breed earlier than their wildland counterparts. Local abundance of granivorous birds—those specializing on seeds, such as chickadees, tits, finches, house sparrows, and blackbirds—increases in proximity to feeders, but this does not appear to translate into abnormally high breeding densities. Black-capped chickadees, for instance, crowd onto feeders and survive better over the winter by doing so. Yet come spring, the *deeeee-deeeee* song of males apportions the breeding population to well-defended and typically sized territories. Those birds that survive the winter but fail to obtain a breeding territory likely filter into less optimal habitats where they queue for better territories and provide an important reserve should breeder numbers in prime real estate crash.

The calories in supplemental foods, especially those from fat and protein, allow birds to emerge from winter in better condition than birds reliant on the natural foods available during a northern winter. The condition of a bird in the spring may directly enhance reproductive output. In Flagstaff, Arizona, pinyon jays flock to feeders for oily sunflower seeds and peanuts. This behavior often enables the "town flock" to lay an extra egg relative to flocks that relied only on the boom-and-bust production of native pine seeds. In Ireland, researchers tested this idea by hanging peanut feeders in forty-five forests and monitoring the reproduction of resident blue tits. After six weeks and about

Cat with European robin

145 pounds of nuts, the researchers were heroes to the blue tits. Peanut-fed tits laid eggs earlier and fledged almost one extra chick per nest than did nearby tits in forests lacking feeders.

Breeding earlier and producing more offspring may seem like a surefire recipe for success, but this is not necessarily the case. Young birds typically dine on insects that become abundant later in the summer. Thus in some situations early fledged chicks may experience tough times, unless they, too, head to the feeder. In subirdia this is really the least of their concerns. Cats are much more troubling to mother birds and their offspring.

In cities, birds collide with buildings, crash into windows, become disoriented by night lighting, and succumb to pesticides. But adding up all the estimated loss of birds from these sources does not even come close to the estimated 1.4 to 3.7 *billion* birds that cats kill *each year* in the contiguous United States alone. Cats in Canada kill another 196 million birds. Considering that Americans own 84 million cats and tolerate another 30 to 80 million feral cats that range freely across our lands, these numbers aren't that surprising. To a bird, however, they are simply horrifying. In Bristol, England, cats eat nearly half of all house sparrows, dunnocks, and robins each year. One in every ten birds in the United States will see the same thing just before death: a cat.

We now know that the effect of cats goes well beyond killing. English researchers demonstrated that a parent bird reduces the rate at which it feeds nestlings when it glimpses a cat. Reduced feeding may result in undersized young, but more important, the unattended nest becomes easy prey for jays and crows.

Some well-meaning residents of subirdia spay and neuter feral cats but allow them to live outside. Others declaw or defang their pet cats. Many hang bells around their cats' necks or try bibs that foil a cat's predatory pounce. All such attempts to limit the effects of cats on native birds are ineffective. Stewarding our urban birds requires that we keep cats indoors. Period. (It's safer for the cats as well.)

I could see right away that the spotted towhee entangled in the net was banded. He was also a few inches away from freedom and gaining on it quickly. I dashed at him and cut off his escape route, gently folding the soft net around him and securing him in the "bander's grip." I cradled his head between my index and middle fingers while controlling his wings and feet by encircling his body with my palm and other fingers. His deep red eye peered at me from his ebony head while his seed-cracking, conical bill tried in vain to split my knuckles. My full attention was on his two legs. I hoped he was the bird I'd seen here throughout the year and suspected to be the long-standing territorial male. The left leg sported two plastic rings: one olive, one powder blue. On the right was a dull aluminum one. Its loss had prevented me from a definitive identification, but etched in the aluminum was a unique number that could solve my mystery. The etching was faint: 842-29538. Yes! This was the old male I had suspected. Apparently, his black-and-white-striped band had broken off during the past winter. I replaced it, upgraded his other bracelets, weighed and measured him, and let him go back to the business of being a towhee. He'd have a big year ahead; as the summer wore on I noted he had a new mate and successfully fledged two broods.

Though weighing less that a few spare coins and finding food by scratching among the fallen leaves for bugs and grains throughout the year in drizzly Seattle, the towhee I just released was at least seven years old. When I first captured and banded him five years before, he was in adult plumage, so at least two years old. At that time he lived on the edge of the forest. Now that forest was a series of wooded alleys and walking trails interspersed among busy streets, barking dogs, and the general hub-bub that goes along with a new fifty-home subdivision. Although his world changed, the towhee stayed and reigned as avian king of this patch of Earth. I would enjoy seeing him continue to do so for another two years, finally disappearing at the age of nine or more years. When he vanished, probably dying, as do all old animals, he would be as old as any other bird in our study. During the seven seasons I

Spotted towhee in bander's grip

watched him, he had four different mates and sired at least ten fledglings from five successful nesting attempts. I banded one of his sons in 2007, but like most young towhees, he was never seen again. Long life, it seemed, was one pathway leading to sustainability of suburban bird populations.

To estimate the lifespan of a bird, we must catch, mark, and resight individuals. We do this by deploying fine "mist" nets throughout an area, waiting for flying birds to bumble into them, as the old towhee had done. To increase our capture rates we also lure territory owners into the net by playing an intruder's song. Intent on expelling the interloper, the defending adults fly toward the speaker and sometimes into the net. Some birds become quite entangled, so our first duty is to carefully remove each bird from the net and place it in a soft, dark, cotton bag to calm and constrain. We weigh each bird in the bag and then reach in to grip and extract it for additional measurements—beak dimensions, wing length, and tail length—and to note signs of past injury, breeding condition, parasitic flies and mites, and overall health. Finally, we band the bird's legs with a unique sequence of plastic color and metal numbered rings that allow us to quickly identify each individual with binoculars. We record all this information in field notebooks and later enter it into computer databases. If we recapture a bird, such as the towhee in 2006, we measure everything again and record these data. The senior towhee actually was 20 percent heavier in 2006 than when I first tagged him in 2002; perhaps his weight increase was a consequence of a new nearby bird feeder or an abundance of bugs and berries. It probably helped him through his only doubly productive year. If we only resight the banded bird, as I did with the towhee in 2007 and 2008, the color bands allow us to note where the bird currently lives, confirm its survival, and, with further observation, document its reproductive output. Netting and trying to resight banded birds year after year allows us to estimate a bird's lifespan and put annual reproduction into a lifetime perspective.

The lifespan of a small songbird in subirdia is surprisingly long. The majority of birds live less than one year, but some live much longer. The longevity record set by the towhee of at least nine years was closely followed by other towhees that lived at least six or seven years. One song sparrow matched the old towhee, living to be at least nine years old. We also banded Pacific wrens and Swainson's thrushes that lived at least seven and eight years, respectively. The oldest juncos, Bewick's wrens, and robins we encountered were at least six. Most of these record holders lived either in forest reserves or in the places we studied during their transition from forest to neighborhood. In these places, the adult birds we recaptured—meaning that they were already at least two years old when we banded them—lived an additional two years on their territory. Apparently, though some birds move during construction, others stay put—for life.[1]

Keeping longevity records is enjoyable; nothing beats a day in the field when I glimpse a bird I've watched for several years. When I do, my history with that individual floods forth as I recall past nests, mates, offspring, travels, and tribulations. I feel as if I really do *know* another creature. But to know whether a population of birds is sustainable—able to persist on its own accord without the need for input from populations that live elsewhere—I would have to determine the likelihood that the average adult and the average fledgling would survive the coming year. A little algebra helps.

We usually do the math for the average female and try to determine whether she is able to fledge at least one daughter that in turn lives long enough to fledge a daughter of her own, and so on. Such production would keep the number of mother birds—and by extension all birds—in a population constant from year to year. The number of birds in next year's population equals the survival of this year's adults, plus the number of young they produce discounted by their ability to survive the coming year. So, for a population to remain constant—the ratio of this year's population to next year's equal to 1—requires that a mother's survival rate plus her annual production of fledglings multiplied by their annual survival rate equals at least 1. For a mother Pacific wren that produces only 0.7 females per year on average in a neighborhood under construction, replacing herself requires that she and her fledgling both have a 59 percent chance of living to breed another year. That is unlikely. But for a mother junco who produces 1.7 females per year in the same construction zone, to replace herself requires only that she and her fledgling have a 37 percent chance of living another year. For a small bird, that is more than reasonable. Let's see whether the birds of subirdia can hit that mark.

Dave Oleyar is a rare breed, a person who is equally adept in both the wild and the computer lab. For his dissertation research, he determined the sustainability of our marked birds, beginning with an estimation of survivorship in subirdia. This work required him to determine the probability that a marked bird is both alive and observed each year. For example, when we encountered a particular towhee in 2002 and again in 2004, we knew it was alive though unobserved in 2003. By considering each bird's encounter history—the year-by-year record of whether we saw the bird or not—Dave calculated the chances that an average adult and fledgling bird of each species in each type of landscape we studied lived to see another year.

Because adults move little after they establish a territory, Dave thought our estimates of their survival were pretty good. By repeatedly checking on territories, we rarely missed a bird one year that later turned up alive. As a grand average, Dave calculated that just over half of all adult birds survived each year. It mattered little what type of bird he analyzed. Robins had the highest survival—more than seven in ten survived each year in forest reserves. Wrens, both Bewick's and Pacific, had the lowest annual rates of survival, around 50 percent. The chances of survival also varied modestly between birds inhabiting forest reserves, lands undergoing development, and established neighborhoods. We discovered that two classic adapters, the song sparrow and dark-eyed junco, were most likely to survive the year in developments. Abundant food, which these species readily exploit at feeders, is likely a key to their longevity in subirdia. In contrast, classic avoiders, such as the Swainson's thrush and Pacific wren, survived best in forested sites, as expected. The loss of nesting habitat and native foods in developments is costly to these birds. The mathematics for towhees, robins, and Bewick's wrens surprised us a bit, because these species survived best in reserves, rather than in developments. These common residents of subirdia may best be thought of as tenuous adapters that, while able to persist in most settings, thrive only where there is a good mix of forested and built environments.

Dave's machinations suggested that most adult songbirds have annual rates of survivorship adequate to balance their productivity in some of the settings that characterize subirdia. But as encouraging as these numbers were, his assessment of fledgling survival was discouraging. For a population to be sustainable, adult survival must balance reproductive output discounted by the survival of young birds throughout their first year. Because many young birds disperse, often over considerable distances, distinguishing disappearances during the first year that are due to dispersal from those that actually represent death is daunting. Most ornithologists consider it the "black hole." Virtually every estimate of survival before breeding conflates death with dispersal and as a result underestimates survival. Dave's estimates were no exception.

Catching young birds during their first summer of life is not so hard. We caught and banded seven hundred. The problem is finding them again. We adequately covered their home turf the next year, spent some time looking in nearby likely places, and let local citizens know that if they saw a banded bird, we'd be interested in getting a closer look. With all of this effort, we found only ten of every hundred banded. When Dave analyzed these dismal encounter histories, he estimated that roughly 20 percent of song sparrows, spotted towhees, Bewick's wrens, and Swainson's thrushes survived their first year. It was even worse for juncos and Pacific wrens; they just vanished from most settings. Robins seemed to do better; one in three survived, at least in reserves. If these estimates were even close to accurate, then few of the populations we studied were sustainable. We needed another portal into the black hole.

Kara Whittaker held the young song sparrow gingerly. I'm not sure who—bird or student—was more nervous; both had good reason to be. As Kara restrained the struggling fledgling, I helped her slip an elastic loop over each of the bird's legs and position a tiny electronic fanny pack just above the bird's rump. This miniature radio transmitter weighed just over half a gram—less than the weight of twenty grains of rice—and would silently tick every few seconds for three weeks. Kara would use a radio receiver dialed into the transmitter's frequency to find the sparrow and make daily records of its behavior, chart its location, verify that it survived the day or locate its remains, and measure its habitat. After three weeks the radio tag's battery would die, leaving us again in the dark about the bird's fate. Even later, the cotton thread holding the harness together would rot, freeing the bird of its high-tech butt pack. We watched the bird briefly in a box before letting it go. Its movement was normal, the transmitter looked fine, and we received a strong signal from it on our radio. We returned our electric sparrow to its parents and looked for another bird to tag. Over the course of two years, Kara caught and radio tagged 122 newly fledged birds. She got over her nerves; the young birds did not.

Radio tagging is a definitive, although labor-intensive, way to determine a bird's survival. Each day that we find the bird, we know whether it is alive or dead. Some days, a bird is AWOL, but mostly this occurs when a bird leaves the study area, not when it dies. It seems we rarely lose track of dead birds; I've pulled them out of irrigation pipes, dug them out of badger dens and vegetable gardens, isolated them in hawk nests, and extracted them from the gullet of a great-horned owl. (Okay, I didn't pull it from the owl's throat. Another of my students studying young prairie falcons tracked one transmitter for several days that was nearly fully swallowed by an owl until the predator coughed up a pellet that included the functioning transmitter and a young falcon's bones.)

Three weeks go by fast, but it was the best glimpse into a young sparrow's life that technology could provide. Transmitters suitable for wrens and juncos would last even less time, so we decided not to track them. Kara had a longer window into the world of young Swainson's thrushes, towhees, and robins—six, eight, and nine weeks, respectively.

Young birds are tough. Only 20 percent of radio-tagged birds died during our study. Birds such as Cooper's hawks and mammals such as Townsend's

Fledgling robin with radio transmitter

chipmunks, ermine, and Douglas squirrels were the most likely predators. The most notorious of all bird predators, the out-of-the-house cat, was implicated in only one death, though we could never be entirely sure which mammal or which bird had killed the fledgling. (The rarity of cat predation in our study is courtesy of the coyote, a frequent resident of our neighborhoods that has a fondness for felines.) Deaths of young birds, just as we had found with adults, occurred most frequently in the forest and rarely in the surrounding yards. We saw no evidence of young birds being run over by cars, shot by kids, or killed on impact with windows.

The young survivors did not wait for the grim reaper. They moved out. One recently fledged robin traveled more than half a mile in a single day. It must have hopped and walked most of that distance! Robins and Swainson's thrushes were extremely mobile, averaging movements of one to four city blocks per day. Their mobility likely reflects their migratory tendencies and reliance on distant, patchy foods such as fruiting shrubs and concentrations of insects. Robins in particular sought out lawns and pastures that were rich in earthworms. Young towhees and song sparrows made a beeline for the nearest bird feeders, typically covering less than a block a day.

Although the vast majority of young birds lived to see the next day, we would expect fewer than half to live a full year. Extrapolating from the probability of surviving one day to the probability of surviving a full year means that we assume the chances of death remain constant as a bird gains experience, encounters its first winter, migrates, and searches for a breeding territory. That assumption is certainly not true, but it gives us a starting point. And for song sparrows, Swainson's thrushes, and towhees, it provides a more realistic estimate of survival than did our approach of banding and trying to resight young birds. Kara estimated that nearly half of sparrows and thrushes and a third of towhees survived their first year. The other two dozen studies that have tried to estimate survival during a bird's first year found similar rates. Suburban bird populations with juvenile survival nudging 50 percent could be sustained.

Young robins are an important source of food for most everything that has teeth or talons and crawls, runs, or flies through the forest. If we extend the 50 percent mortality rate during our nine-week-long study to a full year, we would expect less than 1 percent of young robins to survive. Our banding study showed that this prediction is incorrect; as difficult as they are to find a year later, we found one of every three robins born and banded in subirdia. I suspect that the high mortality Kara observed is fairly typical of a robin's first few weeks; they are plump, conspicuous, and clumsy just out of the nest. Those that survive this vulnerable time gather in great flocks that cut their risk of predation and increase their ability to track the ebb and flow of food. This social behavior likely lessens mortality considerably relative to their early life, invalidating our naïve extension of early daily mortality rates to the full year.

Using our most realistic estimates, it seems that survival of young birds approaches what is necessary to sustain our most productive breeding populations. The nearly 50 percent annual survival of song sparrows would balance their high annual productivity. Slightly lower survival of young robins, Swainson's thrushes, and towhees may be adequate in the settings where adult survival is highest or productivity exceptional. We remain less certain about wrens and juncos, because transmitter studies were not feasible and our limited ability to find the young birds we banded suggested unreasonably low survivorship across most lands. Nonetheless, we have estimates of the three main demographic rates—adult survival, juvenile survival, and annual productivity—that allows us to more formally appraise sustainability.

Dave made a stew of suburban bird vital rates—the essentials of survivorship and reproduction—and cooked them with linear algebra to estimate the expected annual change in population size year after year, something demographers call "lambda." He did this with the average survivorship and productivity values as well as with values reasonably expected in the best of times. The latter helped account for the fact that we know we underestimated production and survival of young birds. He even bumped up first-year survival with Kara's projections. A lambda value of at least 1 indicates a stable population able to maintain itself with reproduction. Populations characterized by a lower lambda value are often called "sinks," because they would dwindle to extinction over time without replenishment from other places, known as "sources."

Changing lands, where forest is cleared, streets paved, and houses built for suburban families, are at best moderate sinks for birds. Under the most favorable demographic conditions, robins, Bewick's wrens, and song sparrows may be sustained in these areas, but Pacific wrens, towhees, and Swainson's thrushes appear unsustainable in this rapidly changing world.

There are flurries of success here—exceptional reproduction by sparrows and juncos, for instance. But mortality or movement is greater than reproduction by most species, so their populations decline during the construction of subirdia.

Established developments provide the resources that juncos and song sparrows require to be sustainable. Bewick's wrens also perform best here, and our highest estimates of their vital rates suggest they will prosper. Pacific wrens fledge many young in developments, but extremely low survivorship quickly eliminates them from this environment.

Reserves of forest in the sea of suburban development likely hold the key to sustaining many of the birds that live in subirdia. Robins, towhees, Pacific wrens, and Swainson's thrushes are sustainable when breeding there. The young produced in these bits of nature wander widely, feeding on the bounty of suburban neighborhoods. This ability to nest away from residents but to visit their gardens and bird feeders increases the survival of adults and especially their teen-aged offspring. For avoiders such as Swainson's thrushes and Pacific wrens, forest reserves are critical. But they are also important to the persistence of some adapters, such as robins and towhees that harvest the best of the natural and built worlds, which together form subirdia.

Amanda Rodewald and her students at The Ohio State University reached a similar conclusion. They banded nearly two hundred Acadian flycatchers and scrutinized their survival and annual productivity. These small, slightly crested migrants are classic urban avoiders that, like our Swainson's thrushes and Pacific wrens, also frequent urban settings. Sustainable populations in nearby rural areas guarantee their presence in subirdia. It is in these wilder places that flycatchers settle first, breed longest, and produce the most fledglings. The flycatchers in subirdia are mostly young and small—backups to the big leaguers that live in forest reserves. Suburban lands alone are adequate for many adapters and probably most exploiters. Rodewald found this to be the case for cardinals. But to enhance subirdia's riches with species that may survive, but only rarely thrive, functional connections with nearby reserves are required.

Observing the actual connections among populations of birds is in practice difficult. Kara Whittaker's radio telemetry research gave us a first glimpse. In the future, new genetic methods hold promise to advance her insights to help us uncover past movements recorded in the DNA of the birds present today. In theory, by comparing the genetic signature of an individual with the average signature of various populations, one can determine whether a particular bird is most like those it lives among or most like those from a different location. The former would suggest that the bird is of local origin, and the latter would suggest that it is an immigrant. In practice, there is much uncertainty, especially considering the high mobility of birds within a city. Thomas Unfried likes a good challenge, so he pricked the wing veins of hundreds of sparrows and analyzed their genetic code, looking for likely relatives in distant places. He found some, even at sites separated by miles of urban land. And he detected a pattern among our study sites where sparrows were more likely to emigrate from productive source populations and immigrate into less productive sinks. Song sparrow sources—in our study the developed lands of subirdia—not only provided vibrant local populations, they stocked the more challenging places where local sparrow populations would decline to extinction if left alone. Connectivity was especially important for these marginal sinks.

The potential to sustain birds in subirdia is no guarantee of continued vibrancy. Reproduction and survivorship vary tremendously from year to year depending on climate, food, predator habits, and luck. This random fluctuation in demography is often the last nail in a species' coffin. It is why we no longer have ivory-billed woodpeckers, heath hens, passenger pigeons, and seaside sparrows among us. Even the sustainable vital rates we observed for wrens, towhees, thrushes, and sparrows are fragile. These animals all live in relatively small populations; fewer than a hundred pairs inhabit a typical square half mile of a neighborhood. Modeling expected animal fluctuations in such populations suggests that few will persist on their own accord for more than a decade. As long as population crashes are not widespread, one neighborhood's towhees or sparrows can be rescued by production and dispersal from another neighborhood. This occurs in other birds living beyond suburbia, such as acorn woodpeckers.

Documenting the movement of birds that connect neighborhoods offers a way for all citizens to contribute to our growing understanding of suburban birds. Many scientists band birds around your homes. It happens throughout Europe and the United States—from Virginia to Minnesota, southern Florida to western Oregon. Because birds are so mobile, a banded one could show up anywhere. So look carefully at the legs of the birds you watch, scrutinize any dead ones you find, and please report any banded bird to the Bird Banding Laboratory, or similar entity. Of the nearly six thousand birds we've banded, only twelve—roughly two of every thousand—were encountered by the public and reported to the Bird Banding Lab. These dozen reports were precious— all were of birds killed during their first two years of life, and most traveled five to fifteen miles after banding. With your help, these valuable data can be increased. You'll even get a nice certificate of appreciation for your effort!

Realizing that a wide variety of birds live in self-sustaining populations in suburbia highlights the conservation value of our neighborhoods. Stewardship of these lands helps keep common and resilient birds abundant. By buffering climatic extremes and providing steady sources of food and water, the places we call home also provide backup breeding reserves for birds that thrive in wilder habitats. To me, the suburb is neither the geography of nowhere nor the ecological desert unworthy of conservation about which I've read. Pieces of paradise have been paved, but it is not entirely a parking lot. What has been created is fully stocked with a wonderful diversity of bird life that is fragile, though sustainable. Subirdia is the geography of life.

Can the same be said of the nearby places where we work and play?

Where We Work and Play

Some view golf courses as islands of green in rapidly urbanizing landscapes, while to others they are a spreading blight of habitat conversion.

—Dan Cristol and Amanda Rodewald (2005)

It is a pleasant, late January morning in the parking lot of my neighborhood Costco. The doors don't open for another two hours, so I'm alone. I did not expect customers, but the absence of the many Brewer's blackbirds that live here is surprising. These work-a-day black and buff birds eat the refuse that shoppers drop in the many-acre parking lot. Iridescent males with pale eyes display from the few ornamental trees that border the parking spaces. They pair with one or two females and crowd their nests together in a colony hidden in the thick, fringing shrubs. They are always here, even venturing inside the store to eat crumbs at the café and get out of inclement weather. I wonder where they are today.

As the time for the store to open approaches, I see the first blackbirds arrive: eight at 9:46, ten at 9:50, and sixteen at 9:51. At 9:55 shoppers arrive and queue at the closed doors. Two male blackbirds join them, waiting! The doors

Facing page: Brewer's blackbirds

101

roll up at 9:57, and in less than a minute three blackbirds are in the store. A few days later I observe the same phenomenon. More than a hundred people eagerly await the store's opening on Super Bowl Sunday, and again the blackbirds arrive just as the people do. Those in the front of the line with the birds notice them and talk about the crafty animals. Two birds fly in as the doors open. The behavior of Brewer's blackbirds is now closely tied to the commercial day; our habits shape their diet and schedule. Their habitat includes the places where we shop and work.

Industrial areas such as the parking lot where I watched blackbirds are not traditionally viewed as important locales for birds. They do not fit my definition of subirdia—the rich mosaic of residential and less disturbed land that harbors tremendous bird diversity. Yet suburban people work here, and therefore these lands are essential in enabling subirdia to persist. Business sites, which include shopping areas, office parks, commercial and industrial portions of the inner city, hospitals, and school campuses, are varied as well. Often they include substantial food and shelter for birds, and they provide unique opportunities to foster bird diversity beyond subirdia.

Don Norman is one of Seattle's top birders. If a bird flits or chips, Don can identify it. In 2004 I was looking to broaden my perspective on birds in residential areas by sampling a wider variety of forest reserves and business sites. My students and I had our hands full, but we needed to put subirdia into context, so I asked Don about his availability. I'm sure my request was a dilemma; I would pay him to do what he loved, but he'd have to bird in some of Seattle's most degraded lands. Don loved the challenge and started contacting businesses. He tried to access lands owned by local municipalities, colleges, and the corporate grounds of Boeing, Microsoft, Aerojet, Medtronics, Boise Cascade, and Oberto. Most were glad to accommodate Don's odd request. In short order we had seventeen new study sites, including business centers, multifamily housing units, and additional forest reserves. Don began to survey these sites using the same point count techniques my students and I used in nearby residential and reserve areas.

Don found eighty-two species of birds, most in the places birders rarely ventured. Caspian terns, peregrine falcons, and belted kingfishers plied the industrial Duwamish River. Common yellowthroats, willow flycatchers, red-winged blackbirds, great blue herons, bald eagles, and marsh wrens lived along the waterfronts and in the ponds and reeds that were frequent within industrial Seattle. Many places were open—gravel, pavement, and bare dirt—yet here Don found a host of early successional species such as the killdeer, white-crowned sparrow, orange-crowned warbler, western meadowlark, and, of course, Brewer's blackbird. Don found seventy of the eighty-two species he discovered in business sites. A third of these were seen only in these sites, including the extremely abundant white-crowned sparrow, glaucous-winged gull, and killdeer as well as rarer tree swallows, green herons, and cliff swallows. When Don finished the summer's work, he reported rather sheepishly that a keen birder could expect to see equal numbers of species in forests of the Cascade Mountains and in Seattle's industrial heart. Though not as rich as residential subirdia, the three forests surveyed yielded an average of twenty-nine species, whereas the twelve business sites averaged thirty-one species. In fact, one-third of the business sites equaled or exceeded the diversity Don observed in the richest forest reserve.

Many of my region's most common birds appear to thrive in commercial and industrial settings. American crows, white-crowned sparrows, American robins, European starlings, and house sparrows were the five most abundant species in the business world. The other members of the fab five—the mallard, Canada goose, and rock pigeon—were regular residents as well. Few urban avoiders were there, but species that adapt to or exploit young forests, brushy fields, meadows, and open grounds were well represented. To me, this is welcome news. Business sites may function as important inholdings to the

Peregrine falcon

common species of subirdia. Redundancy is important, especially in human-dominated environments where unpredictable changes and accidents can quickly put even our most abundant species at risk. Though imperiled species were rare in Seattle's business districts, some local specialties such as the dainty American kestrel and grassland specialists such as the western meadowlark and savannah sparrow were found here and not in the more diverse residential areas of the region. The ability of business sites to provide for open-country specialists is not unique to Seattle.

Half a world away, Dr. Robbert Snep is talking to Dutch business leaders about his bird, butterfly, and amphibian research on their lands. Snep is convinced that conserving biodiversity on business sites is good business. The greenery that provides habitat for animals also allows workers places to relax, recharge, and refocus on the job. Keeping business parks aesthetically appealing to neighbors and engaging in wildlife conservation also enhances the owner's public image. Most important is Snep's belief that providing room for biodiversity at business sites can have important benefits for the birds and other animals that share his city.

Snep analyzed data from volunteer bird watchers who fanned out to document birds of about one hundred sites in business parks, residential areas, and urban parks throughout the whole of the heavily urbanized Netherlands. Over five years, they found 122 species of birds and quickly discovered that, just like Seattle, Dutch suburbia is the place with residences and parks. There they found more than one hundred species of birds! Business sites were not as rich in birds, but they held a sizeable diversity: sixty-six species in all. The birds that used business sites were similar to those we found in Seattle. Most were waterbirds and birds of young, open country. Some appeared to specialize on the features found around businesses. Eurasian oystercatchers, herring gulls, mew gulls, black redstarts, white wagtails, and Eurasian linnets were more abundant in business sites than elsewhere in the city.

That Eurasian linnets were closely associated with business sites was somewhat surprising. This species is declining in parts of its range, including The Netherlands. Linnets are not alone in being a species of conservation concern among working people. Twelve other birds on Europe's "red list"— terns, skylarks, pipits, and partridges among them—were also found on business sites. The linnet's rosy breast and forehead remind me of its close relative, the common redpoll. This delight adds a splash of color to my otherwise gray winter workdays. Surely, linnets similarly buoy Dutch workers' spirits. It is reassuring to realize the reverse is equally true. The places people work buoy the likelihood that some birds of conservation concern will survive our urban world.

Belgian businesses do more than just provide space for birds. Some are actively collaborating with conservation groups to enhance their lands for rare plants, butterflies, and toads. The Port of Antwerp, for example, has agreed to set aside a large area inhabited by orchids, bats, birds, and the rare natterjack toad. The toads bask on open ground, which is frequently created on port lands but also often disturbed. Setting aside a stable reserve for the toads away from the growing portion of the port development has the potential to double the current toad population, provide business leaders certainty that their future actions would not be of concern to conservation groups, and demonstrate that commercial and conservation interests need not always conflict.

Look around your place of work and the businesses you support. Odds are you will find pieces of nature surviving in the rough edges, manicured hedges, open lots, rooftops, ponds, and puddles that remain. Nearly every commercial area provides habitat for birds and other species by default. Some, such as the Port of Antwerp, are engaged in intentional stewardship. This blend of passive and active conservation benefits nature and provides interesting diversions for workers and customers who might even find a hungry blackbird waiting with them for a shop door to open.

Substantial and significant open space exists within a city's workplaces, but this is not the only green area in the city that can serve double duty as a place for people and birds. Soccer pitches, football fields, and baseball diamonds provide loafing areas and, if the turf is natural, feeding areas for some birds. But by far and away the recreational site within a city that has the greatest potential to double as bird habitat is the golf course.

Golf courses are one of the most plentiful vegetated spaces within suburdia. Worldwide there are more than 31,500 courses, and their numbers are increasing rapidly. In the United States alone 300 courses were built annually over the past three decades. From a humble beginning in Scotland, golf courses now cover nearly 1 percent of the United Kingdom. Courses also dominate recreational space in Australia, Japan, Southeast Asia, and, increasingly, China.

Red-headed woodpecker

The typical eighteen-hole golf course occupies 133 acres of land, only half of which is manicured greens, fairways, and tee boxes. The remaining out-of-play area includes rough grass, shrubs, water features, forests, and other places of possible use to birds. These areas have the potential to replace some of the natural habitats lost from urban areas. For example, since the 1700s half of all U.S. wetlands have been drained and filled, then plowed or paved. Could the water hazards that challenge golfers mitigate some of the lost wetlands that are so important to birds? Many researchers are trying to answer such questions.

107

The first hole on Balmoral Golf Course in Battle Lake, Minnesota, is a straightaway, four-hundred-yard-long par four. I waited for the twosome on the tee to hit their drives so I could lag behind and bird without disrupting their play. I stuck to the rough, which included patches of oak and aspen savannah, marshes, and pinewoods. I got the birdie I sought just short of the green on my first hole. The muffled *waaamp, waaaaamp, waahhhmp* led me to a thirty-foot-tall, broken-top, scraggly aspen. On the tree's northeast side I could see an anxious velvet head poking out of a small hole. A female red-headed woodpecker, her headdress glowing in an early ray of sun, cried out in hunger. Her mate soon arrived and fed her. I watched him hawk insects in the rough and ferry his catch back to his incubating partner as she sat tightly in the dark nest cavity they drilled deep into the soft heart of the old aspen tree. His uniform was British—deep red, black, and white. But this bird is uniquely American, one of our more sensitive species that has declined substantially throughout its midwestern range.

Red-headed woodpeckers are often found on golf courses because these settings imitate the oak savannahs that the species naturally lived in. Open, moderately disturbed habitats have declined as oak, hickory, and beech forests matured without the diversifying effects of natural fire. Other places this species used were lost as borders with trees around row crops were plowed and orchards were cut. Golf courses have taken up some of the slack. Where some dead trees and limbs remain, woodpeckers nest as successfully on courses as off of them. In Ohio, three of four woodpecker pairs fledged at least one young, regardless of being in the company of golfers. The pair I found in Minnesota was well along the same path.

The red-headed woodpecker was only the start of a great round of birding at Balmoral. In the nearby conifers bordering the sixth fairway I got an eagle. A bald eagle, no less! From only a few feet above me a majestic adult, its white head and tail stunning against the green landscape, flew down the adjacent fairway. Play stopped as all the golfers marveled at the national bird with whom they shared space. As I headed to the clubhouse, I racked up chipping sparrows, American goldfinches, American robins, and mourning doves from the grassy understory and red-bellied, pileated, and downy woodpeckers from the trees. Associated with the many woodpeckers were those that lived on the results of the woodpeckers' industrious lifestyles—secondary cavity nesters, including eastern bluebirds, tree swallows, eastern pewees, and white-breasted nuthatches. When finally a scissor-tailed flycatcher floated overhead, its ten-inch-long tail plumes dancing in the air, I figured it was time to sign my scorecard and call it a day. The golf pro wasn't impressed with my score, but he was tolerant and pleased to have indulged my strange request to bird his course.

I asked the pro whether he did anything to promote birds at Balmoral, and he answered "no." He knew some birds needed dead trees, but that was not permitted. He wanted aesthetically pleasing, live trees. Weeds, even the dandelions so favored by the goldfinches, were constantly fought and eliminated. Fortunately for the birds, and the players who clearly enjoyed sharing the course with them, this requirement seemed not to matter. Although the pro focused on the fairways and greens, the birds thrived in the rough. Here there was room for people and birds and their mutual interaction.

Balmoral is not alone in simultaneously providing recreational and bird habitat. Nearly everywhere, golf courses increase the local diversity of birds. In southeast Queensland, Australia, golf links support an average of 450 individual birds of more than one hundred species, especially waterbirds. Parrots, such as the ubiquitous rainbow lorikeet and abundant sulfur-crested cockatoo and crimson rosella, are also common there. In Quebec, Canada, courses equal the local diversity found in nearby natural parks. British, South African, and Italian courses all support bird diversity exceeding nearby suburban and agricultural settings. Throughout the United States, courses support forty to eighty bird species, including many of my favorites, such as the anhinga, little blue heron, and brown thrasher. Waterbirds thrive on the many ponds that challenge golfers; in Florida more than ten thousand individuals of forty-two species live on a typical course.

Though golf courses offer some mitigation for the global loss of natural lands, they foster mostly common urban exploiters and adapters. Few birds of conservation concern thrive on the typical course. Local rarities, such as the brown-headed nuthatch, cerulean warbler, blue grosbeak, alder flycatcher, winter wren, and blue-winged warbler, are lost from most courses. But where native vegetation is maintained, forests are conserved, and natural buffers around courses are developed, rare species coexist with golfers. Red-headed woodpeckers exist where large, decadent, and native nut trees are retained. Natural grass and shrublands support loggerhead shrikes, American kestrels, upland sandpipers, and burrowing owls. Ponds and streams connected to functional watersheds and surrounded by shallows rich in native aquatic and emergent plants provide foraging, loafing, and nesting habitat for native moorhens, ducks, waders, and kingfishers. In arid regions courses offer an exceptional ability to provide diverse riparian areas. In New Mexico, for example, courses rich in streamside vegetation attracted sixty-five species of birds not found in the surrounding desert scrub, while excluding only seven dry-land species.

In contrast to the diversifying effects of golf courses built in already developed areas, the opposite often occurs where rich native lands are cleared. When a few hundred acres of native coastal "Strandveld" in South Africa were converted to a golf course, for example, an estimated eighty-five hundred individual birds were displaced, and four species—the grey-winged francolin, black-shouldered kite, Cape grassbird, and pied crow—were driven locally extinct. Though the combination of converted land and remnant patches of Strandveld, roughly in a mix of three to one, actually held a higher number of species than occurred before development, the resulting mix of birds was dominated by a few common species and many rare ones. On the estate, pigeons and doves became common, as did a few wetland and open-field species, such as the cape wagtail, blacksmith lapwing, spotted thick-knee, and the beautiful pin-tailed whydah. Their local gain came at a regional expense of the four lost Strandveld specialists.

The birds of St. Lucia paid an even higher price. Only around fifteen hundred white-breasted thrashers exist in the world, all of them on the two tiny Caribbean Islands of St. Lucia and Martinique. Travel and tourism drive these island economies, so the pace of resort construction is furious. As a result, nine of every ten bits of native bird habitat have been cleared from St. Lucia. Native bird populations are reduced to small, endangered capsules crowded into the few remaining fragments of nature that survive. Each successive resort chips away at the tenuous existence of the birds. Le Paradis resort was built in 2006 on top of the largest remaining thrasher population. It took a quarter of the birds' remaining habitat and cut the world's population by 15 to 20 percent. Large developments on small, biologically rich islands are incompatible with nature conservation. Even the best attempts to limit destruction and mitigate effects are destined to fail—there is quite simply not enough room for the inevitable miscalculation of humans and randomness of nature. The greens fees of island golf are priced beyond what nature can afford to pay.

Sometimes golf courses make real contributions to conservation. A beloved electric blue, rust, and white songbird was recovered from the brink of extinction in part because of nest boxes installed on golf courses. The eastern bluebird population in the United States and Canada declined from the 1930s to 1970s, eventually being reduced by an estimated 90 percent. Clearing of nesting habitat, overzealous use of pesticides, and a series of very cold winters transformed a favorite and common bird into a species of conservation concern. Cavities in dead trees and branches drilled by woodpeckers or forged by natural decay required by bluebirds and other secondary cavity nesters disappeared during this time as hedgerows were cleared for agriculture and forests were eliminated or neatened for housing developments. Suppression of fire allowed remaining forests to thicken and crowd out the open glades where bluebirds hawk and pounce on their insect prey. At the nadir of the bluebird population, in the late 1970s, concerned citizens established "bluebird trails" by placing thousands of nest boxes along fence lines, roadways, and golf fairways where open forests remained. The birds responded quickly, increasing steadily by about 3 percent on standard survey routes from 1980 to 2011. Boxes on golf courses were readily used; in most cases the number of eggs laid and hatched, the condition of chicks, the number of fledglings reared, and the survival of fledglings equaled that at boxes elsewhere. A critical ingredient in successful nesting is the presence of nearby cover, which allows young fledglings to hide from predators, such as sharp-shinned and Cooper's hawks. By emulating the naturally disturbed settings that bluebirds prefer, golf courses augmented with nest boxes are making important contributions to the restoration of a desired bird to the woodlands exploited by people.

If the bluebird is a sign of happy coexistence between golfer and bird, the goose is just the opposite. Geese flock to courses, nesting and swimming in ponds and grazing the grassy fairways and greens. Where geese graze, they crap in prodigious amounts. Their waste fertilizes the soil, but before the poop degrades, it is messy green goo that clogs a golfer's spikes and infuriates most players and residents of golf developments. Geese are also quite aggressive and stand their ground as golfers come near; some birds chase and bite people or bash them with their stout wings.

In South Africa, Egyptian geese are on the rise and frequently live on golf courses, such as Cape Town's Steenberg Golf Estate. More than eight of ten estate residents view geese as a problem, and most call for a severe culling of the flock. In a recent survey, one-quarter of all golfers felt that all of the geese, a native species, should be removed. But efforts to harass, cull, and relocate geese are expensive, unsettling to some people, and must be ongoing to be even moderately effective. Most managers see these actions as being ineffective in the long run.

A more sustainable approach to reducing the use of golf courses by geese is to make the space less appealing to them. Increasing distances from open water to grass, reducing the turfed area, and employing border collies to chase geese are effective ways to reduce the environment's attractiveness to grazing birds. The very nature of a golf course will always include some prime goose habitat and therefore will always attract some of these birds, but planting natural shrubs rather than turf where it is not necessary will benefit other species while also limiting geese. Such has been the experience of Scott Nelson, a course manager in the sandhills of south-central Kansas.

The natural links at Prairie Dunes Country Club in Hutchinson, Kansas, are a birder's and golfer's delight. The course is challenging and scenic, with fairways that undulate up and over sand dunes. A poorly hit ball is always in a tough lie, deep in the native grass and scrub rough. The native prairie birds and people of Hutchinson both win from the dunes' development. The course has hosted many national championships, adding to the town's economy. The natural vegetation of the course itself has lower maintenance costs than would be incurred on a well-watered, all-turf course. This benefits the owners' pocketbooks and keeps geese at bay. Golfers rarely complain about geese, which are outnumbered by robins thirty to one. Many other birds thrive at Prairie Dunes as well. In a study that compared the diversity of the course with that of a nearby natural area, scientists recorded nearly equal diversity between the sites—fifty-seven species of birds on the course and sixty-three off. Nine species were found only on the dunes, and fifteen were found only in the reserve. Naturalistic golf courses such as this one complement other regional natural areas, including biological reserves, military reservations, greenbelts, farms, and backyards. Together they can provide a rich regional habitat mosaic.

In subirdia, augmenting natural lands by developing recreational areas, such as golf courses, on degraded land can increase biological diversity. When courses have substantial areas of native vegetation and productive wetlands that are not surrounded by dense homes but instead meld seamlessly into other less developed lands, they can become true urban oases. The resort in South Africa, for example, could have reduced isolation of existing natural lands if, instead of being situated in expansive Strandveld, it was built on existing, degraded agricultural or mine lands. Course construction could have included substantial restoration of native species. In these places young golfers become birders, and old golfers on the nineteenth hole talk of birdies on the green and in the trees.

Urban land is often too expensive for most conservation budgets. It is not surprising therefore that nature reserves are rare in the city. Nor should it be surprising that tolerant nature flocks to the fragments of habitat that exist elsewhere in the city. Because of their relatively large size and inclusion of set-aside or lightly used lands, many businesses and recreational sites are de facto urban natural areas. The size of such sites matters to many birds. In Seattle, forest birds exist primarily on larger patches of city forest, those of at least 100 acres. Similarly, in South Africa, patches of remnant Strandveld smaller than 125 acres were unsuitable for most native birds. Conserving areas of such substantial size may be possible only if they simultaneously provide other, traditional monetary returns. Golf courses, business parks, and corporate campuses deliver such revenues and, with some consideration for the use, type, and arrangement of the lands they include, can help conserve common and even

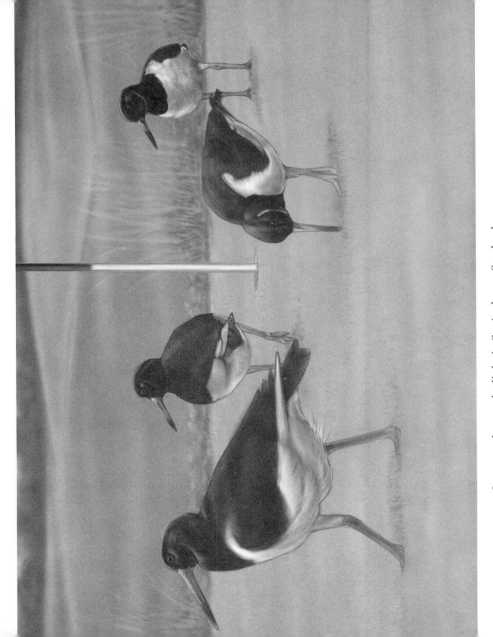

Oystercatchers at the links in St. Andrews, Scotland

some rare species. The birds that enrich our residential areas come not only from the nature beyond the city, but also from the city's working lands. As we use all that the city has to offer, so too, it seems, do the birds.

It may be difficult to conceive of hard-working lands as providing a conservation benefit for birds. Indeed, many caveats need to be considered. Toxic chemicals from industry, expansive banks of windows in office parks, traffic, mowing, grading, clearing, and even well-intentioned nature watchers challenge birds. These challenges can be met first through increased awareness and knowledge. Paying attention to the riches that lie outside the cubicle, boardroom, truck cab, and clubhouse may compel those of us inside to be better stewards outside. Knowledge empowers us to avoid sensitive areas during critical breeding, feeding, or resting times, lobby for the use of native plants, and work with city planners and neighbors to advance ecologically complementary land uses—an idea that favors siting similar lands near one another. Cooperation among neighboring landowners is at the heart of making conservation effective in the city, and this is especially true in its most industrial core, where green spaces are rare and typically scattered. These bits of open space around warehouses, ports, and rail yards, or the acres of lawn that flank office parks and professional buildings, could become significant to birds and other forms of life if they were managed collectively rather than individually. Collecting small but adjoining lightly used lands into a larger area devoted to conservation could offer animals the space they need and allow like-minded landowners to share the costs and benefits of providing for them.

The ability of some birds to adapt to the most severe challenges of city life is indeed impressive. In Scotland, oystercatchers seem to outnumber golfers at the very birthplace of the sport, St. Andrews. The black-and-white waders with gaudy orange beaks roam the links in huge flocks, plucking grubs from the turf and resting away from the cold Atlantic surf. Our actions incidentally provide what some birds, such as oystercatchers, require. Our actions can also directly sculpt a species to fit better within our world. Cliff swallows nest by the thousands under highway bridges. As the birds fly in and out of their nesting colonies, speeding commuters kill many. Despite the slaughter, the birds persist in large numbers. In fact, highway deaths are on the decline, in part because the swallows are evolving more agile bodies. Surviving swallows have longer and thinner wings than those hit by cars, a feature that endows them with greater, life-saving maneuverability. Nature is fragile, but it does not always break when bent: sometimes it evolves.

SEVEN

The Junco's Tail

Walk out of your door and find some evolution.

—Steven Palumbi, *The Evolution Explosion* (2001)

In subirdia a diverse community of birds lives in a robust, if fragile, ecological web that intertwines humans as both producers that facilitate the presence of others and consumers that threaten the web's basic integrity. As birds struggle to survive and reproduce within the web, they do what all of life does: evolve by the process of natural selection. This evolution can adjust a bird's physique to better live among us, and it can fashion its behavior. Natural selection sorts among the genes that underlie these traits, but it also can work directly on cultural traditions passed through generations by social learning rather than genetic inheritance. Both forms of evolution—genetic and cultural—occur in sentient and social species such as birds. That evolution happens and that it adapts birds to our challenges should come as no surprise. After all, Charles Darwin drew heavily upon the human-directed evolution of pigeons to bolster his theory. But the lightning-fast pace of evolution in subirdia, I suspect, would take even Darwin aback.

Facing page: Dark-eyed juncos

The dark-eyed junco I held lived more than a century ago in the forests of Labrador. The small, black-and-white "snowbird," as the species is often called, was shot in 1891 for science. Its collector, likely L. M. Turner from the Smithsonian Institution, expertly skinned and stuffed the sparrow-sized animal, preserving its outward anatomy—part of what an evolutionary scientist would call the "phenotype"—for future biologists such as myself to inspect. I carefully spread its tail, which consists of twelve feathers, six on the left mirroring six on the right of the bird's midline.

The junco's snowy outer tail feathers contrast strongly with the charcoal inner ones. As juncos flit between bushes, this white flag is conspicuous, but it is quickly furled out of sight when they land. The sudden disappearing act may confuse a predator in pursuit, while during flight the white signal may serve to organize the escaping flock. Both sexes have white edges to their tails, but males have the most, and they flash this signal to would-be rivals as they defend breeding and feeding sites. Aggressive males chase one another among the bushes and through the airspace with their small tails spread out, revealing the white, like poker players laying fans of cards on the table to compare their hands. The junco with the most white wins. The old guy from Labrador was a winner. Each side of his tail held two fully white outer tail feathers and a third that was nearly all white. I scored him a 5.6, meaning that 5.6 of his 12 feathers were white. I've seen only two males with whiter tails.

The bird collection at the University of Washington's Burke Museum has a lot of juncos. Thankful for the large number of specimens, I used them to peer back in history and see how junco tails have changed. Several hundred were arranged before me, belly up and toe-tagged, liked stiffs in a morgue. I reasoned that the male aggression that accompanies a big white tail probably earned my Labrador junco a place here as he pushed a bit too close to the gun long ago. But juncos beyond suburbia may also naturally have bright tails. In San Diego, Dr. Pamela Yeh observed that urban juncos evolved substantially different tails in only twenty years. She suggested that the longer breeding season enjoyed by city juncos favored males that cared for young rather than those that squabbled among themselves for territory. If caring males had fewer white tail feathers than fighting males, juncos with darker tails should become commonplace in the city. I checked the credentials of the Burke Museum juncos and scored their tails to test Yeh's thesis.

Dead juncos talk, and the story they told corroborated Yeh's hypothesis. Junco tails from the wilds—the forests of Oregon, California, the Olympic Peninsula, the peaks of the highest Cascade Mountains, and the north slope of Alaska—averaged 42 percent white, but those from urban western Washington averaged only 35 percent white. Through time, urban junco tails changed little, but it was change in the direction expected; before 1940, when Washington cities were small, tails were 37 percent white, but from 1988 to 2012, when cities were large, tails of male junco specimens were 35 percent white. I put the juncos back in the drawer, excited at the prospect that evolution had occurred in my backyard, and was likely continuing.

Dark-eyed juncos offered an insight into the keys of rapid evolution that we now know is frequent in novel ecosystems such as the city. The ability of colonizing animals to adjust their behavior or appearance to a new situation boosts natural selection in two complementary ways. First, expression of a wide variety of behaviors, such as when city juncos start and end breeding, gives natural selection the raw material it needs to act upon. This is called "phenotypic plasticity." If, in this example, breeding for a long time each year results in more young, then juncos expressing this phenotype will be favored by natural selection. In a novel environment, such as a city, it is likely that phenotypes, rare among colonizers, will suddenly be at a premium. A junco that breeds early in the forest, for example, may be doomed by snow, while one that breeds late into autumn may not be able to fatten sufficiently for migration.

These constraints are relaxed in the warm, feeder-rich city. A new type of junco—one with the behavior of breeding early and often—are favored in the city. If the behavior of breeding all season is inherited by the many young produced, then this trait will quickly come to characterize urban juncos; that is, it will evolve rapidly.

Pamela Yeh suggested that a dozen or so juncos that initially settled in San Diego in the 1980s had capitalized on the extended length of the breeding season. It was likely that these colonizers originated from birds that typically spent the winter in San Diego but bred in nearby mountains where short breeding seasons are the rule. For reasons unknown, some stayed, and of those that did, the ones who bred longest left the most descendants. The juncos that stayed evolved bolder and less easily stressed personalities as well, which allowed them to breed throughout the long nesting season with less interruption. In fact, those juncos that took full advantage of the long summer in San Diego fledged twice as many young as did those that retained their mountain temperaments and breeding habits. Plasticity paid big dividends, allowing the rapid evolution of personality and productivity. This superproductivity was critical in enabling the founders to persist, and this persistence is how the long breeding season gave natural selection a second boost.

Once city juncos expressed their phenotypic plasticity and natural selection favored those that bred repeatedly, then this behavior worked on other aspects of the junco's physique. During the long breeding season, females preferred males that attended to their multiple broods rather than those that quickly left to fight other males. Males also had fewer rivals in the city, because buildings and roads separated junco territories, and bird density was low. Darwin recognized the differences in female preferences and male competition, or "sexual selection," as important and often complementary forces to natural selection. Phenotypes that are advantageous in competition with others of the same sex or in attracting a mate of the opposite sex can evolve by sexual selection. Yeh believed that males that invested heavily in early broods increased the chances that their mates would remain faithful throughout the summer. These males had less white in their tails, and therefore sexual selection muted city junco tails. Natural selection enforced by predators such as cats in the city, which frequently kill ground-feeding and nesting birds such as juncos, may also favor darker tailed, less conspicuous juncos.

Selection, be it sexual or natural, and phenotypic variability are two of the three necessary ingredients required for evolution to occur. Favoring some phenotypes over others will push change, but unless the new trait is inherited across generations in genes or socially learned traditions, it cannot be said to have evolved. Differences could persist between populations without evolution simply by ongoing behavioral adjustment. European blackbirds that live in the city, for example, have dispositions that resemble those of city juncos. They are tamer, more sedentary, breed over a longer period, and recover from stressful situations more rapidly than do shy forest blackbirds. Despite these behavioral distinctions, forest and urban blackbirds are genetically indistinguishable. Not so for juncos. By raising juncos from the city and the forest in a common environment, Pamela Yeh demonstrated a genetic difference. Despite growing up in the same lab, city and mountain birds maintained the plumage differences seen in their parents. Later studies extended these findings to junco personality traits as well. City juncos evolved less white in their tails and an ability to handle stress and to be bold, though tame, because these phenotypic traits are variable, advantageous, and passed down on DNA through the generations.

Juncos are not alone in evolving distinctive looks in the city. Eastern screech-owls, for example, wear either red or gray plumage. Red owls are more common in the city than in the rural forest, possibly because they are less tolerant of cold than are gray owls. Common mynas were introduced to the North Island of New Zealand by bird fanciers and acclimatization societies a century ago. As mynas spread across the island, they diversified in size, becoming smaller and more similar in size between the sexes in southern relative to northern cities. Small size may reflect food scarcity or may be a response to the need for respiring mynas to limit water loss in the cooler and drier south. Regardless of the reason, the rapid change is striking.

House sparrows also quickly evolved size and plumage differences after they were introduced to cities around the world. I could see these differences among the many sparrows at the Burke Museum just as I could measure differences among juncos. Sparrows from the dry steppe of Kazakhstan were diminutive and buffy in color. Those from France and Buryatia, Russia, were similar, with pronounced gray caps and rusty capes—traits missing from our local Washington birds.

Professor Richard Johnston, of the University of Kansas, documented the evolution of racial differences in house sparrows half a century ago. House sparrows were introduced to North America in the 1850s and vary considerably in size, shape, and coloration. As they colonized cities in various regions—cold or dry places such as Detroit, Michigan; Sisseton, South Dakota; and Edmonton, Canada—and warm or humid places such as Phoenix, Arizona; Gainesville, Florida; Honolulu, Hawaii; and Mexico City, Mexico—their physiques evolved. In cold climates, sparrows are large and have relatively short appendages. Large sparrows are known to survive winter storms better than small ones, perhaps owing to larger fat and muscle reserves, better insulation, and an increased ability to retain heat than small and lanky birds. Sparrows are also typically large in dry, western cities where small birds would rapidly lose precious moisture with each breath. In warm climates sparrows are small and have long appendages, allowing them to quickly shed excess heat and keep cool. As is true of many birds that live in humid, coastal settings, the sparrows of Vancouver, Canada, are exceptionally dark. All these differences evolved rapidly, over fewer than fifty years.

Male house sparrows are larger than females throughout their extensive geographic range, probably because of sexual selection for large males during battles to claim breeding territories. The degree to which males are larger than females, however, also increases in northern cities. Exceptional size in male northern sparrows likely reflects the dual benefits accrued during competitive duels over territory and limited food, as well as the ability to better survive severe winter storms.

House sparrows demonstrate a superb ability to track new environmental demands by adapting quickly to new climates. Ironically, despite this great evolutionary potential at places where they have been introduced, house sparrows are declining rapidly in parts of North America, Australia, and especially their native Europe where they have gone from pest to conservation concern in a few decades. Across the whole of the United Kingdom and much of northwestern Europe, sparrow populations declined by more than 50 percent from the early 1970s to the late 1990s. Loss of habitat within cities and suburbs is one important reason for these declines. Affluent homeowners in particular have removed rough grass, nettles, and native weeds that produce the seeds and insects sparrows eat, paved portions of their front gardens to increase parking space, and removed old roof tiles and rotten fascia boards and soffits the birds use for nesting. Sparrows are now mainly found in poor neighborhoods where residents cannot afford such improvements. As sparrows are forced into smaller green areas, they are also more vulnerable to local pollution, competition with pigeons and starlings, and predation by sparrowhawks, tawny owls, and cats. Sparrowhawks, as their name implies, may be especially important. The rise in the abundance and distribution of hawks after environmental pollutants such as DDT were banned closely paralleled the decline in sparrows. These myriad challenges are overwhelming the sparrow's demonstrated ability to adapt and putting the burden of conservation on the urban poor. Nest boxes, bird feeders, increased greening of affluent homes, and restrictions on redeveloping portions of the city—especially places where buildings do not currently exist, so-called brownfield sites—are needed if the sparrow is to once again thrive in its ancestral suburdia.

Rapid, explosive evolution in response to the novel environments humans create is the new normal. The process that adjusts species to this different land has a name: contemporary evolution. It began in earnest during the Industrial Revolution. In the latter half of the eighteenth century, cities such as Manchester, England, were sooty from coal-fired textile mills and coal-burning stoves that heated the homes of around a half million people. Coal's waste—carbonic and sulfuric acids—poisoned the air, sickening humans and killing most aquatic life. People expected to live barely forty years. Only house sparrows, starlings, and pigeons occupied the city center. Industrial England was dark, so natural selection stepped up its pace and darkened the life that remained.

The peppered moth rests motionlessly on the trunks of trees during the day, having flown and foraged all night. Its sole defense from its bird predators is to expertly match the color of the surface upon which it rests. Mostly white with flecks of dark, the most common form of the moth was well camouflaged against preindustrial English trees that stood white as birch, thanks to an encrusting skin of lichen and bryophyte. When coal pollution killed the bright epiphytic crust and blackened the trees with residues of soot and smoke, the white moths were easy meals for starlings, sparrows, and the occasional jay or robin. The occurrence of light-colored moths plummeted, but a melanic form of the species quickly gained prominence. Butterfly collectors documented the result of this natural selection. The population of moths had evolved to match its new urban environment in response to predation. A variety of other moth species as well as spiders, ladybird beetles, bark lice, and pigeons all followed suit.

European jay and peppered moths

England's skies cleared as industry moved beyond coal as a fuel and regulations reduced air pollution. Soot and smoke stains eroded from buildings. Lichens reclaimed tree trucks. Birds found dark morphs of moths more easily than light ones, so again the population of peppered moths adapted. Beginning in the late 1960s dark moths decreased in frequency by about 12 percent per year in some areas, while remaining common in others. Today, this contemporary evolution has restored white peppered moths to Manchester's streets.

Pigeons, in contrast, are still mostly dark. Dark pigeons may have once benefited from camouflage, but for reasons not entirely understood, they also have a slight reproductive advantage over light-colored forms. Their reproductive powers keep them common today despite being conspicuous to predators. In fact, being of a common shade has some advantages. Many raptors actively select the odd bird from a flock of prey even if that bird better matches the environment's color. In cities, this would include the rare pigeon with white markings in a coal-colored urban flock. Regardless of the city's hue, pigeons, it seems, may stay forever dark—a reminder to future generations of the perils of a coal economy.

The soot of coal may no longer be an evolutionary force in most modern cities, but the novelty of the urban environment remains a potent driver of adaptive change. The antibiotics we engineer to fight disease quickly select for resistance in their bacterial targets. Three short years after penicillin was discovered, bacteria were resistant to it. Similar responses to next-generation antibiotics have given us superbugs like MRSA (methicillin-resistant *Staphylococcus aureus* bacteria), which kill tens of thousands of people a year. The pesticides we spray in cities to kill mosquitoes or caterpillars are no match for the heritable variation hidden in insect populations. After the development of DDT in 1939, bugs evolved resistance to it by 1946. Birds unfortunately did not and would continue to lay thin-shelled, easily broken eggs in the United States because of DDT if not for the actions of scientists who successfully sued governments and industries bent on applying it. Even the fish we eat are evolving. Gillnets and sport fishing target large salmon and grayling; small ones squeeze through the mesh of nets and rarely end up on the walls of trophy seekers. As a result, the size of wild pink salmon—an abundant Pacific species—has declined by 30 percent in just four decades. Grayling in Scandinavia have likewise shrunk, closely matching their girth at maturity to the legal net mesh size.

It is easy to see natural selection at work when we are the direct selective agent. We fish; fish evolve. We poison; insects and bacteria evolve. But for every species affected directly, many others respond to a changed community. Removing or adding predators, competitors, or facilitators stretches and warps subirdia's web of life. Sometimes the web responds with splendor, but often it does not.

Hawaii's suburbs contain at best one or two of the spectacular array of native birds that evolved in the absence of humans. Our intended and unintended introductions of predators such as cats and mongooses; competitors such as rats, mynas, and finches; and diseases such as malaria and pox have shredded Hawaii's web. Seventy-five percent of the 125 or so native Hawaiian birds present four thousand years ago when humans first colonized the islands are now extinct. The few that survive will need to evolve disease resistance and the ability to exploit a new diet heavy on exotic fruits. Evolution may not happen fast enough to forestall extinction in the face of such extensive change.

In subirdia, birds not only adjust their genes to the novel environment but also their culture. Birds pit their innovative and flexible behavior against the opportunities (such as new foods) and challenges (such as traffic noise) that are omnipresent in cities. Culture evolves in urban birds if flock mates copy the successful behavioral solutions to these challenges and opportunities. This is the process of social learning—observing and following the actions of others. Learning, the analogue to heredity in genetic evolution, enables actions to wax and wane in frequency as populations adapt. If you thought genetic evolution was fast, hold on! Cultural evolution is even faster; it need not wait for one generation to inherit genes from its parents but can produce change within a generation.

The highway a block from my backyard sounds like a river of moderate size. It surges at rush hour and slows to a trickle at other times, but I can always hear it. Birds hear it as well. It is not just this sort of noise that affects the acoustics of urban environments. Buildings disrupt sound waves and alter sound channels, creating a complex mosaic of loud and quiet spaces throughout the city. To communicate effectively, birds must adjust to the hum of the city; they must avoid the noise, sing louder, or alter the pitch and tempo of their voices. Indeed, social learning and cultural evolution allow them to do all of this.

Urban birds rise earlier than rural birds. The morning songs, or dawn chorus, start sooner in the city than in the country. In Spain, spotted starlings and house sparrows sing earlier on noisy streets than on quiet ones. In the United States, American robins sing earlier, and even well into the night, in cities relative to rural lands. These adjustments in timing reflect the stimulating effect of urban lighting as well as the active avoidance of rush-hour traffic noise.

We talk loudly in noisy places, and birds also raise their voices in the noisy city. Nightingales sing louder on weekdays when urban noise is extreme but quickly hush their voices when the city quiets down on weekends. Noisy miners are noisier near busy Australian streets than near quite ones. These facultative changes in volume likely reflect the energetic costs of singing louder, something birds quickly avoid. Other more subtle adjustments in the tune of urban birdsong are less costly and now known from several species.

The great tit is a signature of European subirdia. Having a bold demeanor and wearing a smart suit of green and yellow with a black vest, these supersized

Noisy miner

chickadees expertly work bird feeders and readily take to nest boxes. They are also models of contemporary, adaptive, cultural evolution that fits their songs into the new urban environment. Loud, low-frequency noise in major European cities favors short, fast, and high-frequency song. In the noisier neighborhoods of Leiden, The Netherlands, tits sing at higher frequencies than they do in quiet ones. Song sparrows in Portland, Oregon, make similar adjustments, raising the minimum frequency of song and directing more energy to the high-frequency components of song in the noisy parts of the city. Even urban juncos trill at higher frequencies than do mountain juncos.

Taste in music is one of the fastest evolving aspects of human life. When I first played "In-A-Gadda-Da-Vida" by Iron Butterfly, my parents were positive the record player was broken. And whenever they played Andy Williams singing, "Moon River," I *wished* the record player were broken. Similar changes in song perception by birds are occurring within our cities. In the Presidio of San Francisco, California, cultural evolution has changed the voices of white-crowned sparrows. From 1969 to 2005, as traffic noise increased, so too did the pitch of sparrow songs. This adaptive response to the masking effects of city noise occurred over time, just as other studies had demonstrated changes in song frequency between urban and rural or noisy and quiet locations. Rather than inheriting the song from their parents' genes, young males in increasingly noisy environments changed their tunes because they could better hear and copy the higher pitched songs of their fathers and neighbors. Low-pitched songs faded into the background and did not stimulate learning. They were also less stimulating to established, territorial males. When scientists dug into their archive of bird sounds and played sparrow songs from the Presidio that were recorded in 1969 to resident sparrows in 2005, the response was tepid. But when sparrows heard the 2005 rendition of territorial song, the modern males lit out in search of the intruder. White-crowned sparrows have about as much appreciation for their ancestors' tunes as my kids have for my parents' music.

Singing soprano may be effective, but it has costs. Though it gets around urban noise, it does not penetrate walls or other urban obstructions as well as low-frequency sound. High frequencies are also perceived as less threatening by other males and may not be readily recognized by potential mates. These miscues could reduce a male's reproductive output, though if females and males continue to evolve in similar acoustic environments, one would expect recognition of a singer's full repertoire to improve. In fact, as urban messages come to differ substantially from rural ones, they may effectively isolate populations and enable the evolution of distinct species. In Australia, silvereyes may be heading down that path. Birds in the city and nearby rural areas differ in song. These differences in language are not yet limiting gene flow, which may explain why urban and rural silvereyes also have no morphological differences, despite singing different songs. Perhaps in time this will change.

A unique aspect of cultural evolution in subirdia is the potential coevolutionary relationship that can develop between birds and people. As our culture changes, we may select for cultural changes in birds. The changed song of birds in response to our changing modes of transportation is a case in point. But this effect is mostly one way, from people to birds. The way people and corvids—birds of the family Corvidae, including jays, crows, rooks, ravens, jackdaws, and magpies—interact leads to frequent and reciprocal, coevolutionary adjustments in our cultures.

Corvids are common components of urban bird communities throughout the world. Interacting with people favors cultural adjustment in these birds because human attitudes and important resources regularly and rapidly change. Three aspects of human culture appear especially important to corvid culture: persecution, provision of new food, and creation of new opportunities. The power of persecution is evident in the nest defense culture of crows and ravens. American crows and common ravens in the western United States aggressively defend their nests in cities and towns where shooting is outlawed and in general where persecution is frowned upon, but they quietly retreat out of gun range in rural areas where an aggressive bird would be wounded or killed. We do not know whether these cultural changes in crows and ravens affect human culture, but annoyance with aggressive city crows is common, and some cities (for example, Lancaster, Pennsylvania) respond with control efforts that may diminish future aggressive tendencies of city crows. In other settings, people bond with crows almost as closely as they bond with their pets, feeding them daily. This practice favors tameness and solicitation by crows that recognize the individuals who provide for them.

City crows seem always to scrutinize us, searching for signs of intent. Dr. Barbara Clucas discovered that they pay particular attention to our eyes. If Clucas looked directly at a crow as she approached it, the crow quickly backed away and even flew off, something that is rare in the city. But if she approached the crow with her gaze averted, the crow carried on as if nothing of consequence had occurred. Casting a stare provides a reliable signal to crows that an approaching human is watching them. If the staring person is familiar, the crow may approach for food. If the staring person is unfamiliar, the crow seems to take the approach as a direct sign of trouble afoot.

The responses of crows to people, be it aggressive or solicitous, has stimulated a radiation of popular culture, from the naming of sports teams and rock bands to the myriad trappings and tales traditional in American Halloween celebrations. Coupled interactions between crows and people in cities include jungle crows scavenging garbage in Tokyo, which prompted improved trash disposal methods by residents and subsequent novel foraging behavior by crows; carrion crows in Sendai, Japan, placing walnuts in roads so that cars crack them open, which encouraged humans to swerve in order to hit the nuts; and tits and black-billed magpies in Britain opening milk bottle lids, causing people concerned about disease transmission to change milk delivery methods. The observed increases in nut-cracking and milk-drinking behaviors have spread slowly from sites of innovation, as expected with social learning and cultural evolution, suggesting that human and bird cultures have coevolved.

The cultural responses of people to nature often depend on the frequency and effect of the interaction, which may be an important factor in the likelihood of coevolution. When birds like corvids are rare, we often ignore or revere them; but once they become competitors, we view them as pests and despise, harass, and perhaps try to control them. Our interaction with urban nature thus has a built-in negative feedback mechanism that favors novel cultures in people and birds as the frequency and type of their interactions change. When ravens were fewer in number and less rivals for resources, for example, we considered them to be birds of the gods, even gods themselves, and useful guardians, navigators for mortals, and efficient sanitation engineers. But when we felt their abundance was reducing valuable game and we became horrified by their consumption of human flesh, we used guns and control policies to reduce their numbers and change their culture. Persecuted ravens became rare and shy around people. This contemporary rareness evoked a sense of mystery, wonder, and concern in enough people that a culture of restoration developed. Understanding our role in such cyclical cultural phenomena may be important to conservation and restoration efforts in urban settings.

Subtle adjustments of bird color and song to urban environments are examples of microevolution. As interesting as the changes are, they do not inevitably lead to the creation of new species—what biologists call "macroevolution." For macroevolution to create a new species, organisms must evolve differences that preclude interbreeding, or their hybrids must be at a distinct disadvantage. When house sparrows colonized Italy nearly four thousand years ago, they injected new genes into the local sparrow gene pool and in so doing started the creative engine of evolution. The result was a new species, the Italian sparrow.

My ancestors crossed the Alps from northern Europe into the lake region of northern Italy during a warm interglacial period around 1600 BCE. House sparrows, commensal with humans at that time for about fifty-five hundred years, likely joined the trek. Once across the Alps, house sparrows spread south and encountered the closely related Spanish sparrow in wilder areas and especially in wetlands and along stream corridors. Probably because house sparrows were rare, especially at the fringes of their distribution, they interbred with the more common Spanish sparrow, creating a hybrid of distinctive coloration; males had the bright white cheeks and rufous cap of the Spaniard paired with the clean breast, small black bib, and gray rump of the northern immigrant. Females of all three varieties are similarly dull and hard to distinguish.

As human populations in the north increased, so did house sparrow numbers. With more pure house sparrow mates available, crossbreeding would have declined. Hybrids, however, were fertile. So as house sparrows kept to themselves and stayed mostly around the Alps, hybrids were likely to selectively pair with one another, learning the proper mate by observing the characteristic plumage of their parents. Hybrids may have been especially forced to mate with their own kind about 350 to 450 years ago when persistent cold weather closed the passes through the Alps for most of each year and curtailed the flow of pure house sparrow genes into Italy. Similarly, the Mediterranean Sea limited the spread of pure Spanish sparrow genes into Italy from southern Europe and Africa. Hybrids became increasingly common throughout much of the Italian peninsula. Today, the Italian sparrow is a recognized species of hybrid origin that interbreeds rarely with house sparrows in the north and remains genetically pure by a combination of its large population size, preference for similar-looking mates, and isolated geography. They are the most common bird species in all of Italy. Some five million to ten million pairs live like house sparrows in all the major Italian cities except those on Sicily, Pantelleria, and Sardinia, where only Spanish sparrows reside.

House

Spanish

Italian

Busts of male house, Spanish, and Italian sparrows

The mobility of birds means that gene flow is typically high among bird populations. As we learned from the Italian sparrow, gene flow and resulting hybridization can be creative, but often gene flow reduces distinctions among populations and prevents speciation, as was true in the case of the European blackbird and Australian silvereye. Even in the extensively urbanized Seattle area, which reduces the mobility of song sparrows and has imparted some slight genetic differences between nearby birds, substantial distinctions are lacking. Here, mobility is an equalizer, but sometimes it is the instigator of change, curtailing gene flow and sending urban and rural populations on the path to speciation.

Blackcaps are small, insectivorous warblers common throughout Europe. For millennia these drab gray birds with sleek black top hats left their German breeding grounds to winter in the pleasant Mediterranean climate of southern Spain, Portugal, and North Africa. It is hard to imagine foregoing a winter in Casablanca for one in the British Isles, but for blackcaps there are several advantages, namely, a shorter migratory route, access to bird feeders, a less variable climate, and a jump-start on the next year's breeding. As a result, what was fifty years ago a rare but normal deviation in migratory direction and distance is now a well-established migratory route. In the 1960s few Brits saw blackcaps at their feeders, but by 2004 nearly one in three feeders hosted a blackcap. This new migratory population orients to the northwest after breeding and is evolving a host of distinctions from the ancestral southwest migrants. And they are doing so despite breeding side by side, a feat rare in animals that is called "sympatric speciation."

Natural selection was able to create a second blackcap that migrates from south-central Europe to the British Isles because it had phenotypic variability to work upon, heritability that connected this variability across generations, advantage for extreme traits shown by the ancestors, and, most important, a way to isolate breeders in one gene pool from those in the other. Heritable phenotypic variation in migration direction and distance provided the foundation for selection because, as with juncos in San Diego, some blackcaps ended up in the novel environment that is British subirdia. At this northern latitude, day length shortens rapidly in the autumn and reaches a minimum less than that experienced by blackcaps on the ancestral southern wintering grounds. This change heightened the birds' responses to rapid increase in day length and exceptionally long days that lay ahead. As spring arrived the British warblers' gonads and migratory restlessness grew more rapidly than did those of birds wintering to the south. As a result of this physiology and a flight plan shorter by three hundred miles, the northern warblers arrived back on the breeding grounds more than two weeks before those using the traditional southern route. Fattened on suet and seed, the northern birds quickly set up territories in the prime locations and selected mates. Because of their early arrival, northern birds paired with other northern birds; few if any southern birds were even on the breeding grounds. This "assortative mating" was the barrier that separated northern and southern gene pools and provided the critical reproductive isolation needed to keep the heritable migratory behaviors of birds wintering in the north and south from mixing each summer.

Natural selection reinforced genetic isolation among blackcaps. If northern and southern birds managed to pair, their hybrid offspring inherited migratory instructions that sent them on an intermediate and deadly flight path. Scientists demonstrated this in the lab with forced pairings; in the autumn, hybrid young oriented west and exhibited just enough flight to end up in the Bay of Biscayne, off the western shore of France. Parents that bred true to their migratory route would produce the most survivors, thus keeping the gene pools pure.

The advantage of arriving first, and perhaps exceptionally fit, on the breeding grounds spurred growth in the proportion of blackcaps migrating north. As more birds rely on nearby subirdia for their winter fare, their features are also changing, just as are those of juncos and house sparrows. Long-distance flights favor birds with slender and pointed wings, so the blackcaps heading to Britain have rounder wings than those traveling farther to Spain. Dietary differences favor distinctively narrower and sharper beaks in northern migrants as well. Blackcaps wintering in the south feed mainly on fruits that require a wide gape to swallow. In the north, a general diet of bugs, seeds, and fat, much of it from bird feeders, favors a narrow bill. Northern and southern blackcaps also differ somewhat in color; the beaks and feathers of northern birds are browner than those of southern birds. This may be a result of differences in molt pattern or simply may reflect the random heritage of the few founders that pioneered the northern route.

The recently and rapidly evolved behavioral and physical differences between northwest-migrating and southwest-migrating blackcaps reflect distinct gene pools despite their coincident breeding grounds. Genetically, these two lineages are more distinct from one another than they are from more geographically distant blackcaps. Separate arrival on the breeding grounds and selection against hybrids allow the unique migratory heritage of blackcaps to persist and enables British subirdia to adorn the northern migrants with round wings and narrow beaks. Subirdia, it seems, not only is capable of sustaining diversity, but also may actually create it.

Northwest-migrating blackcaps have not yet been designated a new species. Systematists—those ornithologists who make such decisions—consider these birds a race or an "ecotype" of *Sylvia atricapilla*. Certainly they are more. They are ecologically and genetically distinct, pursuing a new evolutionary trajectory that is deeply connected to the actions of city people. Many other species are doing likewise. Adjustments in migratory behavior and associated morphology are known in at least fifty species, such as the house finches that live on the East Coast of North America and American crows that inhabit the northern prairies. As the urban tsunami crests, expect many species to settle in deeply with humans. The adjustments they make to our way of life may change course or strengthen, but they are unlikely to recede. Perhaps the next step in blackcap evolution is to become sedentary, year-round residents of Britain. Maybe then they will be officially crowned a new life form. I hope their connection to "people of the city" earns the city blackcap a suitable Latin binomial. *Sylvia urbanus*, anyone?

Birds and other organisms respond to novel environments such as cities in one of three ways. First, many adapt their behaviors, customs, or bodies to the new urban world, and survive. Second, some that exploit cities even change their character so fundamentally that they evolve into entirely new species. Finally, avoiders neither adapt nor transform. Their phenotypes are rigid, unable to offer any possible solution to the challenges of urban life. They become locally or globally extinct. In Seattle, Pacific wrens, Wilson's warblers, and hairy woodpeckers are eliminated from our most urbanized lands because native groundcover and snags are removed. In Edmonton, Canada, the din of the city is claiming western tanagers, least flycatchers, and red-breasted nuthatches. Elsewhere the story repeats. Some combination of changing habitat, novel predators and competitors, disease, human persecution, noise, movement barriers, altered climate and disturbance regimes, pollutants, and toxins stimulate the evolutionary responses of some birds but prove too much for others. Why do some evolve while others do not?

Confronting change as a large population, rather than a small one, is fundamental to adaptive evolution. Extremely small populations, such as the few Pacific wrens that remain in subirdia, are vulnerable to the vagaries of chance and often express little phenotypic variability. Bad luck may affect small populations in some years as extreme weather events kill many or cause widespread reproductive failure. Chance also may deal small populations a skewed sex ratio that effectively limits the pool of breeders to a few paired animals. These sorts of accidental events make small populations even smaller.

Chance events also evolve populations by changing the relative occurrence of variants—what geneticists call "alleles"—through random processes rather than through adaptive adjustment to natural selection. This is called "genetic drift." As a population shrinks, those pairs that survive may be thought of as a random sample of all possible pairs. This random sample will likely include a different frequency of alleles that control important traits than was found in the whole population. This change is evolution caused by genetic drift. As forest was cleared for streets and yards in the subdivisions I studied, some wren territories were removed while others were untouched. Clearing was unplanned with respect to the genetic composition of the wrens; some just happened to be where planners sited streets. The drift in genetic composition of the wrens that remained versus those that were originally present in the larger forest counteracted the population's ability to adapt to local conditions. Perhaps the most innovative wrens or those with beaks or plumage best suited to urban life were removed. The full potential of the wren population was not allowed to confront urbanization. As small random samples of the whole, small populations are less diverse and more under the control of luck than of natural selection. Extinction is often the outcome.

Species that should have an evolutionary advantage in urban environments because of large population size include those that directly benefit from our subsidies, actions, and structures. Species that nest in our buildings, eat our food, and use our planted landscapes should be able to build large populations that are responsive to novel challenges such as exotic predators, new diseases, extreme temperatures, variable food supplies, traffic hazards, pollution, and localized persecution.

High reproductive rates, even in small populations, may also reduce the risk of extinction by increasing the amount of random phenotypic variation within a generation. Species with large broods or that nest frequently each year may increase the production of extreme phenotypes and thereby the chance of rapid adaptive genetic evolution. Eurasian collared doves have been able to colonize and spread throughout European and U.S. cities, even into extreme places such as Alaska, in part because of their ability to produce multiple broods each year.

Short generation times—the average time between the birth of parents and the birth of their young—also reduce the risk of extinction by increasing genetic variability, boosting population growth rates, and increasing a population's ability to respond to natural selection. Generation time is shortest in small animals, so small birds should evolve especially rapidly in urban environments. Evolution might also be generally more rapid in warmer cities where breeding occurs year-round and in species, such as rock pigeons, that have generation times that are shorter than one year.

The social behavior and cognitive abilities of birds also may reduce their risk of extinction. Species that flock and live many years, especially those with large brains relative to their body sizes, often have increased behavioral variability, which can enhance survival during colonization and evolve within a generation by the mechanism of social learning. Advantageous innovations are often visible to flock mates and available for rapid incorporation by a population able to observe and learn or imitate. Corvids, parrots, gulls, and tits may survive and rapidly evolve in urban areas because social learning allows them to develop traditions and avoid random extinction. Birds that learn songs may also rapidly evolve cultural dialects in urban areas that can be important reproductive isolating mechanisms that limit gene flow and enable populations to evolve local adaptations and distinctions.

Isolated populations in novel environments are at risk of extinction but also are primed for rapid evolution. Urban populations that are strongly connected to rural populations may import genes and cultures that may not be adapted to city life. While this variation can be critical to evolution, as we saw when house and Spanish sparrows hybridized to form the Italian sparrow, often it simply slows down adaptation by introducing less fit individuals into the city.

The strength of selection is a fundamental driver of the degree and speed of evolution, so species in particularly extreme, fundamentally different environments, and perhaps under the influence of sexual as well as natural selection, should rapidly evolve or become extinct. Species in newly urbanizing areas or in the center versus the periphery of a city may therefore evolve or become extinct most rapidly. Sexual selection in novel environments may increase the rate of evolution in secondary sexual characteristics, such as was apparently the case for junco tail feathers in San Diego.

From an airplane, cities and towns look like islands. Their shores, more abrupt than fringing reefs, are awash in a sea of agriculture and wildland rather than water. Some are large. Others are small. Many form archipelagos connected by highways. A few are isolated. Birds on these isles of subirdia are evolving, but unlike the radiation of species that is typical of evolution on oceanic islands, we know of only a few new forms that have been created. The ease with which country birds move into cities, while a thrill to urban bird watchers and an essential seed to the evolutionary process within cities, also limits the radiation of diversity.

The evolution of urban birds to their extreme environments has been ongoing for a few thousand years at best. During this time, intentionally or simply as a by-product of our actions, urbanization has given us the rock pigeon, house sparrow, and Italian sparrow. As urban areas become even more extreme, producing entirely new feeding and nesting opportunities and challenging birds with novel soundscapes, landscapes, and birdscapes, the rate of evolution is quickening. Chimney swifts, house finches, barn swallows, black redstarts, and white storks join blackcaps, tree sparrows, and juncos in the march toward urban distinction. The splendid adaptive radiations of island birds have been under way hundreds of thousands of years longer than cities have been evolving birds. Islands in the Hawaiian or Galapagos archipelagos, for example, are one-half to five million years old. It seems certain to me that if birds remain abundant and if cities remain distinctive from the country, many more unique urban bird species will appear. Those are big "ifs."

Keeping cities and neighborhoods within a city distinct will require coordinated urban planning. As the cities of Vancouver in Canada and Portland, Oregon, and Seattle in the United States come to resemble one another and even to meld into a single urban conglomeration, the potential of species that can live with us, such as song sparrows, to diversify is reduced. Maintaining large populations of common birds, setting boundaries to developed land, and planning characteristic city forms are great challenges to a growing urban human population. Meeting these challenges will reward future generations with a diversity of distinctive new urban birds.

An even greater challenge to urban humans will be to ensure that their new urban birds are not alone. Keeping the diversity of birds we already enjoy will require restraint. Large reserves of natural land—prairie, tundra, taiga, forest, desert, rainforest, coastal dunes, mangroves, and all other unique landforms—are required to keep the magnificent diversity of today's avifauna intact. These birds are the raw material for urban evolution. Some colonize, adapt, and evolve. Others restock our cities when extinction occurs. If we are able to retain what we have beyond subirdia and allow natural selection to fashion new species within subirdia, we may begin to balance the great extinction of life we are precipitating. Future generations may come to know us as creators, not only destroyers, of natural biodiversity.

Medicine, agriculture, natural resource extraction, and urban living are strong, contemporary selective forces that together with the ancient actions of island formation, glacial advancement and retreat, and continental drift are shaping today's species, including our own. How we recognize, embrace, and foster contemporary evolution will affect the diversity of subirdia. Confronting the loss of species such as the passenger pigeon, great auk, and ivory-billed woodpecker makes me hungry for new ones. Perhaps I am overly optimistic at our ability to enhance rather than only curtail evolution. When one holds what could be the last individual of a species, or looks into the eyes of a young person stunned at what is being lost from our world, as I do, you grip tightly onto good, albeit faint, news. The life we create does not replace what we have destroyed, but it is important to recognize the positive along with the negative. All destructive forces in nature have a creative side; fires create unique ecosystems and diversify forest structure, earthquakes create lakes, volcanic eruptions enrich the soil. Urbanization too has the ability to create biological diversity.

The partially albino towhee at my bird feeder sports a distinctive coat of black and white. I've nicknamed him the palomino, and I wonder whether his offspring will be similarly plumed when they molt their dusky juvenile feathers. The extra brilliance he wears seems to contradict the role natural selection has played in muting the junco's tail. Yet the expression, and the eventual sorting among random mutations, is also part of the creative evolutionary process. The palomino reminds me of the changes seen in foxes that Russian scientists have bred to be tame during the past century. With a change in fox personality came barking; prick ears, a raised tail, play, and a diversity of coat colors. Could albinism in urban birds also be an unintended consequence of adaptation to yard life? Aberrant coloration is common in backyard birds. Considering such possibilities brings us close to the evolutionary process. Time will show me whether predators weed out the palomino's genes or whether those genes spread. I am happy to be able to watch the action from a front-row seat. Evolution is not something that happens only in the abstract eons of geological time. It is happening right now, in your own backyards. You are a force, as potent as the glaciers of the last ice age, shaping your feathered neighbors.

EIGHT. *Beyond Birds*

Urban sprawl is a quiet killer.

—Mark Southerland and Scott Stranko,
Urban Herpetology (2008)

It is not surprising that birds can handle many of the challenges of living with people, given that they have wings to propel them and an ancient lineage, including dinosaurs, as ancestors. Add the engaging personalities, flexible behaviors, and short generation times and birds are a recipe sure to succeed in even the grittiest parts of our human-dominated world. We've seen this success play out as a great variety of birds interact and evolve in our cities, and especially in the outlying areas that comingle built and natural areas into a true subirdia. But what about the other beasts whose origins are more humble and that must crawl and slither over the built and developed environment? Do those that challenge, frighten, and threaten us, or require pure water, air, or soil, also thrive in cities? My suburban yard has hosted black bears, coyotes, and bobcats, but a comparison of mammal sightings with bird lists in the city will quickly persuade you that most mammals are less able than birds to live among us. And while the occasional alligator lizard and garter snake in

Facing page:: Flying foxes camping in Melbourne, Australia

my garage startle me, amphibians and reptiles rarely thrive in our presence. Insects are more complicated; some that, like birds, can fly and others with rapid evolutionary rates indeed share subirdia.

The birds of Yellowstone National Park and New York City's Central Park have impressed me with their similarities. Both parks host specialists and generalists, predators and prey, seed and insect eaters, and those requiring deadwood and downed wood. The mammals I have encountered in these two localities, however, have been vastly different. In Yellowstone I have seen a dozen species, including the entire hoofed assemblage native to the western United States and most of their predators. Bison, pronghorn antelope, bighorn sheep, and elk graze together in mixed herds on the flats above the Gardiner River. The lowlands are filled with mule and white-tailed deer. Coyotes, ears cocked, pounce through the last bits of snow, trying to crash in on the voles that traveled unseen on subnivian runways. Foxes seek mice and scavenge with coyotes on the elk and bison killed by wolves and winter. Grizzly and black bears mostly sleep, still hibernating out of sight during my late-winter visit. I have seen none of these animals in Manhattan. Coyotes are rare but present in Central Park. Sometimes true exotics are seen, such as the white Bengal tiger that strolled through in 2004. But on my days in the park the only mammals I saw were humans, horses, dogs, and eastern gray squirrels. Raccoons, opossums, and rabbits may have been watching me from under cover, but even these urban exploiters would find it difficult to survive in the crowded park. Mice and rats surely were there, and possibly were abundant, with few nocturnal predators. The message from my brief visits was clear: Yellowstone's mammal community rivaled that of Africa's Serengeti, but Central Park's was a mess.

Mountain beaver

In Seattle, we have studied the response of mammals to urbanization more systematically. Roarke Donnelly and I enlisted two undergraduates, Amy Jennings and Karin Hoffman, to help us survey the mammals in thirty-five of our bird study sites. We worked from the urban core to the Cascade wilds in woody patches ranging in size from one to several thousands of acres. In each we established a grid of small live traps, aluminum boxes baited with oats that caught mice, voles, rats, and weasels alive. We also set out track plates, a pair of sooted aluminum squares that captured footprints of medium-sized animals attracted to a can of cat food we placed between the two, and camera traps, motion-sensitive cameras that recorded sneaky animals attracted to a treat of meat and fish scraps. We trapped small mammals for three nights and monitored the track plates and cameras for six days. Our efforts confirmed the presence of two dozen mammal species, from black bears and spotted skunks to Pacific jumping mice and the odd shrew-mole.

As with our bird studies, we discovered that the variety of mammals peaked in suburban and exurban landscapes where forest patches were moderately sized (50–150 acres). The diversity of the ground vegetation seemed to enhance mammal diversity, so that in residential settings even substantial patches of forest (up to 2,500 acres large) maintained rich communities, as did small (5-acre) forests in the sparsely peopled exurbs. In extensive forest beyond subirdia, the variety of mammals we caught was a bit lower, and in all but the largest urban forest patches it was substantially lower.

Although the form of the relationship between urbanization and mammal variety paralleled that which we found for birds, the composition of mammal and bird communities was distinctly different. We observed a mix of native bird species from young and mature forests and from grassland, tundra, and meadow, which increased suburban diversity, whereas the mammals of subirdia drew their variety from nonnative species. There were six species of native mouse, shrew, weasel, squirrel, and shrew-mole and an equal number that were invaders: opossum, cat, dog, eastern gray squirrel, raccoon, and rat. Urban mammal communities also consisted equally of native and nonnative faunas. In contrast, the mammals of large and wild forests were nearly exclusively native species, with only one nonnative, the ubiquitous eastern gray squirrel.

The small mammals of Seattle respond to urbanization a bit differently from the mammals of Ohio and Argentina. In Ohio, diversity also peaked in suburban settings—neighborhoods and golf courses—but it was composed almost exclusively of native mammals, including shrews, voles, mice, raccoons, and opossums. In contrast to the findings from Seattle and Ohio, rodent diversity steadily declined in Argentina from a rich costal forest reserve through parklands and shantytowns to the industrial center of Buenos Aires. There, nonnative species also buoyed urban mammal diversity, as in Seattle, but even to a more extreme degree. In the city, no native species were found—only house mice, Norway rats, and black rats.

Checking live traps in the morning is one of my favorite jobs. Like Christmas morning, one never knows what the shiny packages will contain. A closed trap door suggests that something is inside, and as I lift up the gift, I know when it's something special. A heavy box is likely to hold a sleek weasel. My heart races as I push in the door to sneak a peek, and then my brain starts to figure out how to get the weasel out of the small box and back into the brush rather than up my pants leg. A light one probably holds a mouse that will need to be weighed, marked, and released. When I checked the twenty-four traps I set out in early August 2000, most were closed, and all were light. The number of white-footed mice, commonly called deer mice, that I caught was staggering; thirteen the first day. A second day produced similar results: six new mice and seven that I had caught and marked the previous day. The third and last day yielded four new and nine known mice. My traps covered an area about the size of an American football field's end zone within a small suburban forest patch a tenth of an acre in size. The twenty-three mice I caught there meant that in each area about the size of a standard hotel room, one could expect to find a mouse. No wonder coyotes and bobcats often surprised me during early morning bird counts! Many other mammals that exploit a city's abundant food and shelter reach densities on par with that of the white-footed mice I trapped.

In Lafayette Square, which is adjacent to the White House, every acre holds twenty or more eastern gray squirrels—ten to twenty times that of rural woodlots—because food and nest boxes are provided. The density of raccoons in suburban Cincinnati approaches that of people (1.7 versus 2.7 per acre, respectively). Shelter provided in the neighborhoods—church attics, abandoned buildings, and sewers—bolster raccoon numbers well above those of typical woodlands. In southern California's Santa Monica Mountains near the subdivisions of Los Angeles, every one hundred acres supports a coyote, a tenfold increase over more distant, less developed areas. California coyotes near people supplement a natural diet of mice and rabbits with trash, livestock offal, and fruit; in Seattle they add cats to the menu. Other moderately sized omnivores, such as badgers in the United Kingdom, skunks and bears in the United States, hyenas in South Africa, dingoes in Australia, and wolves, foxes, and raccoon dogs in Europe, also likely reach extreme densities in our well-stocked cities and suburbs.

The warmth of cities also attracts some tolerant mammals. Foxes and coyotes often seek warm den sites on cleared slopes or where asphalt radiates its acquired solar heat to the surrounding land. Large, nomadic fruit bats take it one step farther. Southern Australia is no place for a big bat that thrives in the humidity and heat of more tropical northern latitudes. In Melbourne, however, irrigation has increased humidity, while buildings and pavement have increased temperatures. Both changes are to the liking of grey-headed flying foxes. As this nomadic bat wandered, some discovered Melbourne's Royal Botanic Gardens, where lush vegetation was a perfect substitute for their typical tropical forest farther north, so they camped in large roosts, numbering upwards of twenty thousand. So abundant were they that by 1986 they damaged a number of rare, heritage trees. (If you've ever looked a flying fox in the eye, you'd know that it wasn't just the tree damage that caught the attention of Melbourne park staff. It was also the stench of several thousand two-pound bats, the males of which secrete pungent oil to mark their territory and attract mates. With the warming effects of climate change, we might expect more tropical visitors to high-latitude cities, with odiferous results.)

The diversity of mammals in a typical city is much lower than the diversity of birds, and while they respond to urbanization in much the same fashion—some adapt or exploit, while others avoid—the relative proportions and notably the place of origin of the animals in these categories is quite different. The majority of native mammals are urban avoiders, whereas the majority of adapters and exploiters are nonnative and often cosmopolitan species, such as the Norway rat, house mouse, and eastern gray squirrel. There are notable exceptions where native mammals, such as flying foxes, European badgers, and hedgehogs colonize and thrive in urban settings. The response of cougars, however, is more typical.

Cougars, also known as mountain lions or pumas, once lived across North and South America. Today, they live mostly in the wilder portions of the continents, although increasingly they intersect the human world where subdivisions spread into native forest or chaparral in the western United States and Canada. Cougars are large and powerful cats that are capable of killing large prey, some up to five times larger than themselves. As developments creep into cougar habitat, the cats continue to use the natural habitat around settled lands that hold their prey—mostly deer. Some occasionally prey on livestock and come face to face with people in surprising places. Such conflicts, which are most common in the exurbs, inevitably lead to cougar deaths. Cats that repeatedly interact with people are often selectively killed. Others are hit by cars, shot for no reason, or contract disease—notably feline leukemia—from domestic cats and die. The result is that surviving cougars rarely spend more than a fifth of their time in subirdia, and most stay far from it. For these reasons most other native large carnivores, such as wolves and tigers, are also urban avoiders.

That some mammals thrive while most others decline in urban landscapes suggests that, as with birds, the makeup of mammal communities will transition in the face of new development. Indeed, as the chaparral outside of Los Angeles is developed, native mammals that utilize woody vegetation, such as the pinyon mouse and desert woodrat, give way to more typical grassland in-habitants, such as the California meadow mouse, harvest mouse, and Pacific kangaroo rat.

Where outright changes in species occurrence do not result, the barriers to movement inherent in a city, such as freeways, reservoirs, and channeled rivers, often affect the genetic composition of mammals that remain. Coyotes and bobcats live on either side of the Ventura Highway that cuts through Los Angeles. Though many animals cross the freeway, few from the east side of the road successfully breed on the west side, and vice versa. Dispersers have difficulty breaking into the breeding population because when animals cross the road, they immediately encounter a dense wall of breeding territory own-ers; territories seem especially to pack up against either side of the highway. The result is that the road disrupts gene flow among the coyotes and bob-cats of Los Angeles, and slight genetic distinctions between east and west are apparent.

Amphibians and reptiles, or what we science geeks call "herps" (from "herpe-tology," which is the study of amphibians and reptiles), fare even worse than native mammals in urban settings. Around the world, the diversity of these animals plummets in the city. Losses of native herps are not balanced by gains in nonnative, urban adapters. There are few examples of apparently benign introductions of anoles, geckos, and lizards in cities and luckily just as few devastating examples of nonnative snakes invading sparsely inhabited ecosys-tems, such as the limestone forests of Guam or the Everglades of Florida. Most of the missing herps of suburdia are snakes and salamanders; some lizards, turtles, and frogs are usually retained. A few tolerant ones—northern brown snakes, bullfrogs, and red-eared sliders in the eastern United States; oblong and snake-necked turtles, carpet pythons, and blue-tongued lizards in Australia; sand lizards in Berlin; and the marsh frog throughout Europe, for instance—become abundant.

In a few instances the collapse of urban reptile and amphibian populations is well documented. New York City supported fifteen species of snakes and salamanders in the late 1800s, including poisonous copperheads and timber rattlesnakes. Today, only two snakes (neither poisonous) and three salamanders (all rare) survive. They share the city with five species of frogs, two species of toads, and six species of turtles. The Italian wall lizard is a new arrival firmly established in the city, and occasional boas, cobras, and alligators escape their cages and scare residents. A richer complement of snakes, salamanders, toads, frogs, and turtles lives just beyond Manhattan in the wilder parks of Staten Island and Long Island. London's natterjack toads were lost in the 1960s, and poisonous adders and common toads have been greatly reduced. One-quarter of San Francisco's herps were extinguished from the city between 1855 and 1963, and today, although sixteen native species can be seen, only four—the rubber boa, coast garter snake, California slender salamander, and arboreal salamander—are considered common. In just half a century, from 1920 to 1973, all species of salamanders, five species of frogs, and two species of snakes were extirpated from Hamilton, near Toronto, Canada.

In the state of Maryland, scientists wade the streams every year to systematically search for amphibians and measure the chemistry and composition of the watersheds that affect them. By checking more than three thousand locations over a decade (1995–2004), they found an astonishing diversity of stream and moist, riparian forest amphibians—forty-one species in all, about half frogs and half salamanders. They also discovered that urban sprawl, and especially the pavement associated with it, was the biggest stressor to amphibians, lowering their diversity precipitously. Worldwide, urbanization affects the quality of the waters salamanders and frogs use by scouring streambeds with flashes of dirty storm-water runoff; polluting streams with deadly chemicals, including chloride from road salts; and enriching them with nitrate from fertilizers and decomposing yard and animal waste that spurs algal growth and its inevitable decomposition, robbing the stream of its lifeblood—oxygen. These factors acted together in Maryland to extinguish most of the diverse stream salamander group from reaches where more than a third of their watershed was paved. As a result, the highly urbanized corridor from Baltimore to Washington, D.C., has lost most of its stream-dwelling salamanders and many of its frogs since the mid-1970s.

Seattle has a rich amphibian community that also has suffered losses as its temperate rainforest has been dissected and dried by settlement. Oregon spotted frogs and western toads no longer exist in the Emerald City. Gone too from most fast-moving urban streams are tailed frogs and Pacific giant salamanders. Northern alligator lizards rarely prowl under driftwood logs or other dry and sandy places along Seattle's Puget Sound. But Seattleites can still enjoy a spring evening concert of Pacific tree frogs and wonder at the sight of dainty long-toed and northwestern salamanders in their moist gardens and forested parks. These species regularly breed in storm-water retention ponds—basins of an acre of so in size that gather runoff and ease its return to urban streams. The rest of the year these animals live in the duff that accumulates under our magnificent forests. In the wilder suburbs of the city, where forests and retention ponds are common, the diversity of native amphibians nearly equals that found in more natural settings; even the occurrence of nonnative and predatory bullfrogs seems to be tolerated by the abundant salamanders, newts, and native frogs. Seattle was never home to poisonous snakes, and its native garter snakes thrive in suburdia, feasting on abundant slugs as well as worms, frogs, salamanders, small fish, and the occasional baby bird.

It is not surprising that amphibians do best in cities that protect and maintain the interface between water and land. This is a crucial transition area that allows many frogs and salamanders to access the wetlands they require for breeding and the woodlands they seek for refuge and food. Protecting wetlands is challenging in urban settings, because land values make it profitable to convert wet areas to building sites by draining and filling. But those that remain are easy to identify and, if the political will is sufficient, possible to retain. Knowing how much upland habitat is needed and how close it must be to the wetland it nurtures is much more difficult. Wood frogs and spotted salamanders disappear from protected ponds in the eastern United States as development fills in around their breeding sites. The same occurs in California, even where threatened tiger salamander breeding ponds are conserved. From these studies and others that have radio tagged and tracked salamanders and frogs as they come and go from breeding pools, we know that the pond itself and the watershed up to about a half mile around it affect amphibians. The entire area need not be set aside, however. Even small animals are able to cross substantial gaps—a distance of a football field or more—between the pond and woods, if these areas include moist grass and some shelter, such as small mammal burrows, leaf piles, downed branches, or native shrubs.

Failing to provide connections between ponds and the uplands that are reserved can be devastating to herps. Common toads throughout Europe are slaughtered as they cross roadways that separate ponds from woods. In France, a toad population was extinguished when every breeder was smashed on a busy roadway that cut across a traditional migration path. Even more robust and resilient reptiles become isolated in parks surrounded by roads. Eastern box turtles in Rock Creek Park in Washington, D.C., for example, are slowly drifting to extinction as new breeders rarely are able to cross the busy roads that border the otherwise significant urban preserve. In contrast,

Spotted salamander

this turtle moves easily among the forests and gardens of suburban Aiken, South Carolina, where fewer and more slowly traveled roads prove to be minor hazards. Still, it is not easy being a turtle in subirdia; some nests are destroyed in yards, and turtles of all ages are inadvertently mowed over or burned in debris piles where they seek refuge. Populations, however, are generally healthy.

Golf courses, with their ponds, out-of-play areas, and reduced traffic, like subdivisions that provide safe travel corridors between breeding and nonbreeding habitats, offer another possibility for amphibians and reptiles to survive in subirdia. In highly urbanized Detroit, Michigan, golf courses support a high diversity of toads, frogs, snakes, and turtles. Suburban courses in Queensland, Australia, on average support seventeen species of reptiles and five of amphibians, including several of conservation concern. Courses in the Sonoran Desert around Tucson, Arizona, also are used by nearly all the species of lizard, snake, and tortoise found in nearby, less disturbed deserts. Many species are extremely abundant on the natural areas included on courses; even some rattlers hunt the rough. In New England, spotted salamanders, as well as wood frogs and American toads, successfully migrate across golf fairways between breeding ponds and uplands. Where golf courses provide connections between natural uplands and wetlands, amphibian populations may be sustained. These connections appear to erode through time, especially if development encroaches from outside the course. As a result, older courses may be less useful to herps than more recent ones.

As critical as connectivity is for amphibians, the type of wetland—whether seasonal or permanent—is also important. Permanent ponds and lakes, which are the most common types on golf courses, often harbor fish and bullfrogs that eat adult, and especially larval, amphibians. Predator-free, seasonal wetlands that flood in the spring but dry up late in the summer support breeding and tadpole development. In South Carolina and Georgia, for example, courses with seasonal wetlands sustained nearly all the local amphibians that are unable to coexist with fish, including sensitive species, such as the marbled salamander and eastern narrow-mouthed toad. Nearby courses that only had permanent water bodies held many fish, but less than half of the local amphibian community.

Lizards, like birds, use a variety of habitats and often are more successful than snakes or amphibians at living among us. Lizard and bird diversity in Tucson, Arizona, peaks in suburbs where exotic trees and native desert vegetation are often juxtaposed. There, greater earless lizards cruise the sandy open areas while spiny lizards and tree lizards command the more vertical surfaces. The less developed Sonoran Desert hosts some real beauties—whiptail, zebra-tailed, and horned lizards—but slightly less diversity than subirdia.

Birds, mammals, and herps have charisma. They dazzle us with beauty, song, and mystique. Most get our attention, and some scare us. Relatively speaking, biologists know these animals well, especially those that live among our homes and businesses. The insects, marine creatures, and even many freshwater fish and invertebrates, by contrast, are less well studied and often poorly understood. Humans are landlubbers, so this bias toward knowing those whose lives are most similar to our own is understandable. But Earth is only one-quarter land, and by far and away the majority of life forms are tiny, lack a backbone, and inhabit the aquatic world we find so foreign. Urban ecologists are just beginning to understand how such animals take to the city, but many of their discoveries echo themes we've explored from our vertebrate perspective.

The importance to amphibians of connecting terrestrial and aquatic environments is easy to see, as they literally live in both worlds. Marine and freshwater creatures are, in contrast, wholly aquatic, yet their existence is also closely tied to our actions on land. Major cities of the world are sited on coasts and profoundly influence the seas upon which they rely. Piers, bridges, seawalls, busy ports, and effluents affect the physical structure of the shore, the hydraulics of currents, and the chemistry of water. Globalization of commerce, and the human activities that support it, is responsible for many introductions of nonnative species—from favorite tropical fish to zebra mussels—into ports, harbors, canals, and lakes. The ecological and economic costs of such invasions are devastating. In addition to changing the physical and biological aspects of habitat—sands, reefs, and fringing plants where marine life thrives—many marine species are directly exploited for food and bait. Yet despite these influences, marine systems appear more resilient to urbanization than their terrestrial counterparts, where loss and degradation of living space are more severe. Our freshwater systems may be especially sensitive. Rivers are frequently overworked to transport goods, water crops, provide power, and sanitize our cities. For efficiency and safety, they have been straightened, armored, levied, buried, rerouted, and dammed. Out of laziness or carelessness, they have been overharvested and polluted. Lakes and ponds may be drained in the name of disease management or valued as urban amenities to be treated, tamed, and manicured into submission. Fish and invertebrate diversity typically declines.

The city of Sydney, Australia, is well known for its beautiful harbor. The iconic opera house, with its roof sails mirroring the sails of yachts, seems to symbolize the connection of city and water. But beneath the surface of that water, one can see how the city's development has compromised the harbor's life. The diversity of fishes and invertebrates is stunning but reduced by some 16 percent relative to undeveloped harbors. Seawalls have replaced a more gradual and sandy shore once rimmed with seagrass flats. As a result, seahorse populations have plummeted, many now garnering endangered species protection. The seawall itself supports a rich variety of marine invertebrates that filter microscopic life from the water, graze on algae, or prey on larger fare. Barnacles, tubeworms, oysters, and mussels are here, but the gastropod, whelk, starfish, and crab communities have been simplified. Sea urchins and octopuses have been excluded because the cracks and crevices they require do not exist in the smooth wall. A variety of natural and manufactured reefs provide necessary cover, and where encrusted with mussels and algae, food for fish. Small fish, such as damselfishes and wrasses, abound near pilings and pontoons. Wrecks foster a diversity of larger fish as well. Though affected by the city, the Sydney harbor's marine web of life is embracing its new urban structures.

The fish in North American lakes are less impressed with city life than the mollusks of Sydney. The growth rates on developed lakes of some favorite game fish—bluegill, largemouth bass, and trout—are about one-third of those on undeveloped lakes. The reasons for the growth slowdown are pretty obvious, and carefully worked out by my colleagues Tessa Francis and Daniel Schindler. These limnologists and fish fanatics assessed more than fifty lakes, about half of which were in the Pacific Northwest. They observed that the density of houses ringing lakeshores closely predicted the abundance of dead trees, branches, and other decaying plant growth—what the scientists call "coarse woody debris"—that was found along the shore and in the lake shallows. When the number of dwellings per mile exceeded about six, the coarse woody debris disappeared from the lakes because residents cleared it away for an unimpeded view of the lake and access to its water. Fish growth suffered from the reduced cover and especially from reduced food in the form of insects from the land.

In urban streams, food from the land is also affected by light pollution. Lights on city streets, yards, commercial buildings, and elsewhere attract and kill many urban insects and reduce the emergence of aquatic insects from streams. This double whammy means that insectivorous fish in urban streams have less food and may be stunted in size or outcompeted by species with more general diets. Changes in food as well as increased sewage, silt, nonnative species, and fishing pressure together determine the community of fishes that live among us.

A quick look in the dark corners of one's house will confirm that not all urban invertebrates are fish food. The four million Americans who live in Phoenix, Arizona, know this all too well. Despite a general decline in arthropod diversity—the variety of insects, spiders, and crustaceans—inside the city relative to the surrounding desert, some species are seemingly everywhere. Water is a key feature; it is diverted from the few nearby rivers into the city and put to work. Plants that support arthropods also benefit from the moisture. Wolf spiders, spitting spiders, and poisonous black widow spiders rule the city along with small ground beetles, Argentine ants, fruit flies, European cabbage butterflies, roaches, and termites. Reduced numbers of native scorpions, pseudoscorpions, tarantulas, and large predatory ground beetles are of little relief to most Phoenix residents, whose homes support thirty times the density of black widows found in the native desert. And even though urban female widows lay about one hundred fewer eggs than their desert sisters do, each still produces a brood of nearly two hundred. The shiny black beauties with beguiling red hourglass markings are in Phoenix to stay. In the sunny, warm niches of Swiss cities, it is the diversity rather than the density of arthropods that impresses. There, biologists have identified 163 species of spiders and 139 species of bees (though most bees were just traveling through the city on their way to or from the hive and nectar-rich plantings).

The loss of large predatory insects from urban ecosystems is similar to the loss of large predatory birds, mammals, and herps, as is the increase in some particularly destructive invaders. In Phoenix, Argentine ants outcompete many other native ground-dwelling arthropods, leading to lower overall diversity in places where the ants are abundant. Changes in arthropod diversity ripple up the food chain as herbivory, decomposition, seed dispersal, and pollination are affected.

Despite the abundance of native bees in some Swiss cities, fostering them in urban areas is challenging. In Boston, for instance, bumblebee declines have been linked to excessive road mortality. (Checking your windshield or car grill may give you a firsthand look at this problem.) In Phoenix, bee numbers are also low, raising concerns that the essential pollination services our gardens and crops require may be compromised by the city and its surroundings.

Rot is not something about which many of us worry. But this essential service is provided in large part by invertebrates, and, as a result, it too is

Black widow

affected by urbanization. In the forests of northeastern U.S. cities, leaves decompose quickly, much more quickly than they do in more natural settings. This might seem counterintuitive because city soils are disturbed, of low quality, and polluted. City trees also grow tough, decay-resistant leaves to fight water loss, pollution, and the mouths of herbivorous insects. Soil organisms, however, play a huge role in starting leaf breakdown, notably earthworms, which are more abundant in urban soils than in natural soils. The greater warmth of urban soils is favorable to earthworms and other agents of decay, but the greater abundance of worms in U.S. city soils also reflects their alien origins. Most earthworms are not native to the United States; they came from Europe as settlers brought in plants and horticultural soils from their native homelands.

Some of the most familiar invaders, such as "night crawlers" and "red wigglers," are clearly enhancing soil fertility and even helping stem climate change. As worms decompose leaves, they accelerate the conversion of nitrogen from ammonia to nitrate, making this natural fertilizer available to plants. This conversion spurs plant growth, though it may also favor some quick-growing, nonnative, weedy plants over slower growing natives. Either way, the nitrogen cycling facilitated by worms and other soil microorganisms is an essential service that allows our plants to sustain themselves in the city. Carbon is also quickly stripped from the decaying plant matter by worms and stored in the soil. This may be the worm's greatest gift. Increasing the capacity of our soils to sequester carbon is a significant step that helps counter the carbon we release into the atmosphere. Releasing carbon stored long ago in soils by burning fossil fuels increases global temperatures, so by storing new carbon in today's soils, worms are, in effect, helping us combat global climate change.

To some species the storage of leaf carbon in the soil that worms enhance is occurring too efficiently. Ovenbirds are small warblers that nest on the floor of midwestern and eastern U.S. forests. Their bright song, an increasingly loud *Chertee Chertee Chertee*, is always one of the first to enliven woodlands in spring. Where earthworm invasions are substantial, the breeding output of ovenbirds is low. In part, poor reproduction is due to a reduced ability of ovenbirds to hide their nests on forest soils no longer cloaked in dead leaves. But a reduced abundance of invertebrates, many of which are outcompeted by the invasive worms, also makes it difficult for ovenbirds to capture enough food to feed large broods.

Not all parts of the decomposition network are functioning at top speed in urban systems. The loss of large beetles from cities extends to important decomposers—carrion beetles and dung beetles—at least in New York. In the greater New York City area, fifty species of these busy, burying beetles exist. Only twelve species live within the city, and only four live in Manhattan. Why should we care about these animals that we rarely even notice? Well, they're not called "carrion" or "dung" beetles without good reason! Some scurry about rolling up balls of dung—from horses, humans, dogs, deer, opossums, and maybe even geese—laying their eggs inside the balls and burying them. Others lay their eggs in small animal cadavers and then entomb them beneath the soil. When the beetle numbers are reduced, the dung and the dead fester aboveground, for anyone to experience. As this mess rots and stinks, it attracts flies that fly in for a feast, lay their eggs, and buzz off to visit people and their food.

Not all insects respond as predictably to urbanization as do the beetles. Butterflies are more complex. In Osaka, Japan, butterfly diversity peaks just beyond subirdia, where forests occupy about 70 percent of the land, frequently edging up to meadows, buildings, and gardens. There, an astonishing seventy-eight species of butterflies seek floral rewards. In California and Ohio, Dr. Rob Blair also found butterflies most diverse in suburban, or slightly wilder, settings. But in Singapore, South Africa, Brazil, and the United Kingdom, no peak in butterfly diversity at intermediate levels of urbanization has been detected. Rather, diversity declines steadily from a high in natural areas outside the city to a low in the city center. Much of this variable response can be explained by the occurrence of sugary nectar that butterflies crave. Where nectar-producing plants are abundant and of many types, especially if night lighting is low and traffic slow, butterfly diversity in the city—and probably bee and nectivorous bird diversity as well—will be high.

Evolutionary responses to urbanization are not the sole province of birds and mammals. Insects and amphibians also evolve strategies to urban life. Color changes in response to urban pollution are well documented, as we saw with the case of industrial melanism that darkened much of England's fauna, including many insects, during the coal age (Chapter 7). Moor frogs in urban Russia are also changing their appearance, but not in response to pollution. These hand-sized frogs come in two basic morphs: one with a broad, creamy stripe promi-nently etched down the center of their backs, and one plain-backed without a stripe. Those with stripes mature quickly and therefore survive the ephemeral nature of pond life in hot, dry, and high-elevation environments better than do the plain, slow-growing variety. The ability to quickly metamorphose and hop away from the natal pool is also advantageous in the unpredictable urban land-scape. As a result, Russian cities harbor mostly striped frogs.

Road noise is a factor in the evolution of insect and frog calls, just as it is with birdsongs. Grasshoppers that live along Germany's autobahn overcome the low-frequency background hum of vehicles by chirping at a higher pitch than country hoppers do. In southern Australia, southern brown tree frogs and common eastern froglets croak at higher frequencies near roads than away from them. For frogs, the shift to a higher voice is costly; big frogs with deep voices are more intimidating to their rivals and more attractive to potential mates. The balance between being heard and not being perceived as a wimp must be fine. At present, getting the message out is winning, and pitches are on the rise, just as they are in urban birds, such as great tits, that also balance bravado and clarity. In frogs, however, the change in pitch proceeds much more slowly than it does in birds because while big frogs may be able to raise their voices above the traffic din, this ability is ultimately constrained by size. Big frogs have deep voices because their bodies resonate at lower frequencies while croaking. Change is also slow in roadside grasshopper populations, because it requires either evolution of the nervous system or the toothed comblike structure on the insects' hind legs that are raked against the forewing to "sing." Evolution of physical characteristics, such as body size or leg form, proceeds slowly across generations, but birdsong can evolve culturally through social learning within a generation. Big birds learn to sing soprano much more quickly than big frogs or leggy grasshoppers can evolve new instruments.

Possessing the cultural wherewithal to quickly and persistently adjust behavior to new environmental challenges is one reason that many birds have adapted to cities. An inability to do so is part of the reason why other animals have succeeded less well. Flexibility benefits birds, but much of their success is deeply rooted in their DNA, essentially preadapting them to the fast-paced, urban life.

Wings allow birds to avoid massive road mortality that culls the urban herd of mammals and herps. Grounded on four legs, dispersing animals and migrating amphibians are slaughtered by the millions, especially at night. Snakes, unable to generate body heat internally, as we mammals do, seek roads for the warmth they radiate. On cool evenings they bask on the tarmac and are killed en masse. The full magnitude of roadkill is difficult to estimate, but collisions with large animals are certainly on the rise. In a 2008 report to Congress, the Western Transportation Institute noted that in 2004, one in twenty reported vehicle collisions in the United States—some three hundred thousand—involved wildlife, most often deer. These collisions annually kill two hundred Americans and injure twenty-six thousand, at an estimated cost of more than $8 billion. The cost to wildlife is also extreme. In the United States, between half a million and one million deer are killed each year, and twenty-one species of vertebrates are federally endangered in part because of road mortality; only three are birds. The Humane Society of the United States estimates that more than three hundred million vertebrates die annually in vehicle collisions. Worldwide, the annual death toll is staggering: five million amphibians in Australia, four million birds in the United Kingdom, two million birds and mammals in Canada, one hundred million vertebrates in Spain, twenty thousand to thirty thousand badgers and sixty thousand to eighty thousand hares in Sweden . . . the list goes on.

It's not that birds avoid all cars; it has been estimated that more than eight million are killed annually in Sweden alone. Even birds can't help but collide with cars in the United States, where roads cover 1 percent of the land and stretch for nearly seven million miles. But bird casualties may be responded to—by birds and the people who collide with them—in unique and adaptive ways. Eddie the bald eagle thrilled Seattle commuters by regularly perching on a roadside light. In 2011, Eddie was killed by a head-on collision with a city bus. The public grieved and eulogies were posted, but hearts soared the next year when his mate acquired a new partner and successfully nested. In contrast, few comment on the daily death toll of raccoons, rabbits, opossums, coyotes, deer, and snakes. In fact, we often go out of our way to kill herps, especially snakes.

Persecution is thought to drive snake declines in Nigeria, Pakistan, the West Indies, Russia, Belize, Brazil, Canada, and the United States. Eating frogs may do likewise in Pakistan. Some mammals even draw our wrath, passing a cultural carrying capacity more quickly than birds. It is late summer and my neighbors are battling the moles that burrow under their lawns of turf grass.

Rather than appreciate the services the near-sighted mammals do—tilling and aerating the soil, eating harmful grubs—they see only the messy dirt mounds that multiply each night. Traps are set. Moles die. Lawns remain green and tidy, but there is a hidden cost.

Some animals cause real problems in cities; beavers destroy valuable trees and flood urban streams, overabundant deer harbor ticks that spread disease, cougars and bears occasionally attack humans, and the bite of a black mamba can kill or maim. It is possible to coexist with some of these problems. For example, nonlethal beaver deceivers—devices that limit the amount of water held back by the animals' dams—allow beavers to live safely among us. We can also learn how best to behave around dangerous wildlife and reduce the attractiveness of our yards and neighborhoods to large predators. But sometimes we just need to remove the animal. If a black mamba gets into my house, I'm calling a snake charmer!

Thick skin is an author's armor, but the lack thereof is another reason amphibians struggle in subirdia. Pesticides, herbicides, pharmaceuticals, and salts and other chemicals we use to de-ice roadways readily diffuse through the moist skin of a frog or salamander. These environmental stressors, along with increased exposure to ultraviolet radiation, which has increased as ozone levels have been reduced, interact to lower an amphibian's ability to fight off disease and parasites. Frogs compromised by ultraviolet radiation and pesticide contamination, for example, are less able to fight off infections by parasitic flatworms. The worms, which are superabundant in overly fertile ponds, deform the frogs' legs, often leading to grossly misshapen or extra appendages. Amphibians soak in all manner of environmental challenge through porous skin that evolved to help them live a simple, passive life in a less toxic world.

Less obvious pollutants affect an even wider range of city animals. Our penchant to light the night sky disorients insects and birds, and also confuses reptiles, notably sea turtles. On beaches around the world, when the tide is high and the moon slight, female turtles emerge from the dark sea. On legs built for swimming, they paddle across the sand to the dunes where they use their hind flippers to excavate a niche suitable for one hundred or more leathery eggs. After covering all signs of their actions, the mothers reverse course and return to the sea. About two months later baby turtles dig out of their sandy incubators and orient toward bright horizons. Naturally, this would lead them away from a dark shore and into the sea. But because of shore lighting, the bright horizon today is often an inland town or lighted yard, which leads the hatchlings astray. Well-lit shores may also cause gravid females to delay or abort landings, reducing the use of otherwise healthy nesting habitat. Lighting affects sedentary species as well. Toad tadpoles develop more slowly under the lights because they stay away from illuminated, though productive, shallows. Some diurnal reptiles and amphibians, including a variety of geckos, anoles, skinks, toads, and tree frogs, actually benefit from night lights, which allow them to forage under the streetlamps that are so attractive to insects.

Among the perils of city life that birds and other animals share, the most uniformly serious one is the domestic cat that ventures out-of-doors. In suburban Canberra, Australia, cats eat around a half million small animals each year. Most are nonnative small mammals, more than three hundred thousand per year. Native reptiles are also frequent victims, about thirty thousand per year. Few cats kill amphibians, though upwards of five thousand frogs, toads, and salamanders a year may die by the paw. A similar pattern holds in the United States, where cats kill 6.9 to 20.7 billion small mammals each year. Suddenly the slaughter of more than three hundred million vertebrates—birds, herps, and mammals—on U.S. roads seems tiny. Cats eat at least an order of magnitude more animals than we run over with our cars each year. And it is mostly a different set of animals that cats eat; Fluffy doesn't eat many deer or salamanders, but she really tears up the mice, birds, and small reptiles.

Some herpetologists call urban sprawl the silent killer. Many of its victims are indeed extinguished without notice. To me, however, the city kills with a meow and a squealing of brakes. These sounds are growing, leading city people into small actions that may produce big benefits for the animals that seem capable, yet unable to live in subirdia.

Residents concerned about migrating salamanders shut down Beekman Road, a suburban street in East Brunswick, New Jersey, on February 27, 2013. They do so each spring to give male spotted salamanders safe passage from their upland wintering areas to their breeding ponds. Without their actions, the road would be slick with dead salamanders, and a local treasure would be lost. In cities throughout Europe other citizens are building "toad tunnels" and drift fences to funnel amphibians safely under highways. Elsewhere, signs warn motorists to break for hedgehogs, newts, and all manner of mammal. New streetlights no longer emit bright, multispectral mercury vapor that disrupts animal movements. Owners of swimming pools are also pitching in on behalf of urban wildlife. Pools attract many amphibians that cannot escape the steep-sided, chlorinated water. Pool covers, chlorine-free sanitizers, and "frog logs"—floating mats attached to small, frog-sized exit ramps—are reducing the deadly nature of spas and pools. We are capable of much more.

Subirdia is alive with a great variety of plants and animals. Our actions challenge many, but we live with much more than rats, mice, pigeons, and roaches. The other species include not only the terrifying or pesky. Many share our preference for open ground and grasslands with a sparse canopy of trees. Others take advantage of our natural reserves and recreational facilities. Let's build on the actions of those who live along Beekman Road and take a close look at what else we can do to nurture more and prune fewer branches on the urban tree of life.

We abuse land because we regard it as a commodity belonging to us. When we see land as a community to which we belong, we may begin to use it with love and respect.

—Aldo Leopold, *A Sand County Almanac* (1949)

Steve Humphrey jabs his walking stick into an old farm road, the oaken bridge that once bore the heft of cattle, rotting behind him. I jog my memory to conjure up the road's former course, because it now is invisible. A forest of Steve's making completely consumed it. Just four years ago, when I first visited Steve, his land was a pasture, and the road on which he stands was indeed visible. Today I am amidst a rich mix of young trees, some of substantial stature, and many cloaked with epiphytic bromeliads, mosses, and orchids. Daily toil and a long-term vision are transforming ten acres of Costa Rica back into the cloud forest it once was. Birds—more than 170 species have been sighted, most from the comfort of a fabulous veranda—are singing its praises. As Steve tells my class about the ecology of tropical forest restoration, I watch violet sabrewings sip nectar from *Inga oerstediana*, gaze upon flocks of tanagers seeking the fruits of melastomes, and marvel at the surreal plumage of a

Facing page: Steve Humphrey showcases his restored forest

green honeycreeper and blue dacnis as they slurp ripe bananas from a suspended feeder. (Inga, or colloquially *guabo*, plants are important neotropical members of the mimosa family that produce seeds in variously shaped pods. The melastome plant family is one of the world's most diverse groups, including shrubs, trees, and lianas.) Steve and his wife of forty-two years, Suzie, are good neighbors to a stunning array of life that nudges against the wilderness of Chiripó and La Amistad national parks.

Adjusting his yard to benefit nature is Steve's passion. Unlike the desires most of us pursue, this one did not come with instructions. Instead, much of what Steve practices he has invented. For example, restoring plants to a tropical fen (a small wetland that features a mat of vegetation floating atop water fed by an underground spring) first required that the thick turf mat be killed and then replanted with native sedges. Simply removing the turf would destroy the floating mat's integrity, so Steve and his helpers tried covering the area with black plastic in hopes of suffocating the deep-rooted pasture grass. That strategy worked, and reestablishment of the fen was successful. Today, I watch a northern waterthrush, just back from its breeding grounds in the United States, move among giant prayer plants in a grass-free wetland.

Practice has also taught Steve patience. Rather than directly planting the trees he hopes will eventually constitute his forest, he mimics the natural process of tropical succession. He starts his forest with weedy trees that germinate and grow quickly through pasture grass. As they grow, they shade and kill the grass. With the grass under control, Steve works on the understory plants and slower-growing trees. He disturbs his pioneer forest by girdling prolific melastomes, leaving their skeletons for woodpeckers and allowing light to reach soil he has planted with a great variety of native oaks, cecropias, figs, aguacatillos, tree ferns, and heliconias. These plants come from Steve's workshop—a self-built greenhouse where he tries to germinate seeds and nurture seedlings he has collected. To improve his craft, he experiments with light, temperature, water, and fertilizer regimes to optimize growth and maximize survival.

As a diverse forest begins to take shape, Steve still directs its course. He fells a tree here and another there, in the process creating light gaps for sun-loving understory plants and slash piles for snakes, small mammals, and wrens. The results are self-evident—cool streams in shady ravines (what locals call *quebradas*) line the property that today sports ninety-nine species of native trees. On the other side of Rio Peñas Blancos a mature, protected cloud forest, only slightly more diverse than Steve's yard, eagerly awaits its reoccupation of the former pasture. This rare forest type, which occupies less than 1 percent of Earth's surface, is already broaching the river. Inga trees stretch their branches from the reserve to Steve's property, providing a bridge suitable for monkeys. Soon sloths, tapirs, and jaguars may follow. These and other rainforest species are expected to travel through Steve's canopy as they seek seasonally abundant resources in the highland and lowland reserves that together form the Alexander Skutch Biological Corridor.

As Steve clears, digs, and plants, he dreams about the return of the big mammals. Our reaction is more immediate. My Costa Rican friend Marcos Garcia watches Steve and then whispers to me that he is going to start restoring native forest on his farm. Gerardo, a local man Steve employs, tells me that the hard work of restoration has instilled in him a love of plants and interest in birds. My students also soak up Steve's example, and, after getting dirty planting one hundred saplings, they are eager to carry his message north to their parents and friends. The process of restoration not only benefits nature, but also inspires a respect for the land and what it can do. It motivates us to see property as Aldo Leopold saw it—a living community, not simply a commodity.

To some, Steve may sound fanatical or especially privileged. True, as a retired medical doctor of modest needs, Steve is able to devote his life to the land. Yet in such a situation, few of us would strap on machete and binoculars each morning to battle the tropical heat, briars, and insects on behalf of birds. Fewer still would take the initiative to spread the gospel of restoration to their neighbors, gaining access and financial support to affect land well beyond their ownership. But this is Steve's hobby. Rather than retire to play golf, he retired to practice ecology and conservation. He is making up for an inability to do so earlier in life, when the Vietnam War called him away from his zoological studies. Like golf, he notes that his pastime is expensive, but fun! And like all great pursuits, others enjoy it, too. Downslope on the humid Pacific shore Jack Ewing is restoring Hacienda Barú to include a national wildlife refuge. In the high mountains, Marino Chacón Zuniga and his family nurture ancient oaks along the Savegre River. Both of these former cattle ranchers traded cows for tourists decades ago and now proudly offer extensive, restored native forests for all to enjoy. These are exceptional efforts by unique people in biologically important areas, but even if you live in an apartment in downtown Cincinnati, Ohio, you can be a good neighbor to the wild creatures that share your residence. Let's look at some practical ways to support the native communities where we live.

It's easy to learn what it takes to foster a vibrant ecosystem. The inputs and outputs, so to speak, are quite simple. All life requires food, water, and shelter; if we seek to practice stewardship with urban nature, we must provide these prerequisites. In larger natural systems, solar power and the consistency of creating new soil with ample fertilizer and water for plants deliver all the essentials. Plants in turn are food and shelter either directly or indirectly for nearly all other life, by feeding herbivores that become food for carnivores, for example. In subirdia, and especially in more urban areas, our effect on plants—

removing many and favoring some kinds over others—fundamentally affects the food and shelter nature would provide birds and other animals. So it is here that we must first concern ourselves. But simply provisioning what wildlife scientists refer to as "habitat" in urban settings, while necessary, is insufficient for most life. To persist, animals must be able to live, breed, and move among the habitats we plant and tend. The subtleties required to get habitat right for birds, mammals, fish, reptiles, or amphibians, especially in highly modified or restored lands, are often beyond our initial naïve understanding. As such, the practice of ecology, particularly in dynamic environments influenced by the unpredictable hand of humans, is much more complex than the practice of building a rocket capable of ferrying a person to the moon. Ecologists are constrained by both the laws of physics and the vagaries of human nature. Predicted outcomes are far from certain, although our knowledge is sufficient to offer basic principles to those willing to improve our urban ecosystem for other species.

If the animals that live in our cities, towns, and suburbs could get our attention, I believe they would ask us to consider practicing nine principles, or commandments:

1. Do not covet your neighbor's lawn.
2. Keep your cat indoors!
3. Make your windows more visible to birds that fly near them.
4. Do not light the night sky.
5. Provide food and nest boxes.
6. Do not kill native predators.
7. Foster a diversity of habitats and natural variability within landscapes.
8. Create safe passage across roads and highways.
9. Ensure that there are functional connections between land and water.

These nine commandments from nature provide shelter that is safe and nourishing for the animals that enrich our world and economies. As we have seen, many species grace subirdia, but their persistence is intricately linked to human actions. Actions aligned with these ideals would increase the persistence of biological diversity by increasing the vitality of species that tolerate our presence. Let's take a close look at each of the nine commandments from the perspective of apartment dwellers, small homeowners, large landowners, and those who plan, build, and manage our cities.

Do not covet your neighbor's lawn. Having a "perfect" lawn is an original sin of most Americans. Our love of lawn is rooted in our history as a former British colony, and perhaps even in our evolution on short-grass savannahs, where ancestral hominids found safety from predators. I mowed lawns for a living as a kid. So did my brothers and most of our friends. When we weren't cutting them, we played or relaxed on them. There is nothing wrong with a little lawn. Frederick Law Olmsted, the father of suburbia, espoused the value of lawns as giving his neighborhoods a "sense of ampleness, greenness, and community." Many suburbanites foster lawns to boost the value of their homes, as safe havens for their kids, or as firebreaks. Some see lawns as art, as proof of our domination over nature, or as a way to gain prestige among their neighbors. Whatever the reason, most ecologists agree that the ubiquity of the lawn has outstripped its benefits. Domination of suburbia by lawn constrains the diversity of birds that could be supported. Robins, starlings, crows, wagtails, oystercatchers, and a few other birds forage in lawns, but to my knowledge, not a single species of bird, mammal, reptile, or amphibian reproduces and carries out its other life functions in the modern lawn. It is the ultimate green desert, or worse.

In 2005, 2 percent of the coterminous United States, some forty million acres of land, was lawn. Nearly every bit was "industrial lawn," composed of

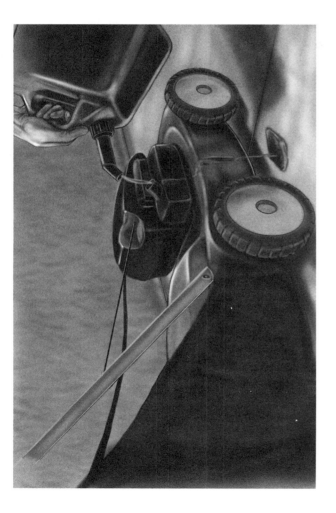

only a few nonnative grass species. These invaders are regularly mowed to a low, even height and kept continuously green and free of weeds and pests. To maintain this sea of grass Americans annually spend $30 billion. They use eight hundred million gallons of gas, seven billion gallons of water, three million tons of nitrogen fertilizer, and thirty thousand tons of pesticide. The use of pesticides alone is ten times greater than used by the average farmer and includes chemicals that disrupt normal hormone function and reproduction, are suspected to cause cancer, and are banned in other countries. Simply filling up gas-powered lawnmowers is an ecological disaster of the highest order; seventeen million gallons of gas are spilled annually. That amount is more than was spilled by the *Exxon Valdez* in 1989 and every twelve years would equal the amount spewed into the Gulf of Mexico during the 2010 *Deepwater Horizon* disaster.

Concern about lawns has sparked a great deal of debate, creative thought, and neighbor-to-neighbor strife. In 1991, a savvy group of graduate students and faculty from Yale University's School of Forestry and Environmental Studies joined their colleagues in the School of Art and Architecture to consider how Americans could redesign their lawns. The resulting book details the history of lawns and charts a plan for those who wish to follow the first commandment. Lawn owners can increase bird use of their turf by reducing its extent, bordering it with shrubs, shading it with trees, mowing it with hand- or electric-powered machines, and skipping the fertilizers and pesticides. Doing this produces what the students and faculty refer to as a "Freedom Lawn." The plant composition of such lawns diversifies into a rich mix of grasses, forbs, and flowers pollinated and grazed by native, beneficial insects, which in turn are eaten by birds and other animals.

The less often a lawn is mowed, the more likely it is to be used by an array of animals. A less-disturbed lawn will attract goldfinches to ripe dandelion seeds, provide nest sites under tussocks for juncos and sparrows, and harbor frogs, turtles, and small mammals such as moles and voles. Spending less time and money on lawn maintenance may allow homeowners to relax and enjoy nature in other ways, such as bird feeding.

But adopting a Freedom Lawn is risky business. Those who do so buck a multinational industry heavily invested in producing seed, sod, fertilizer, pesticide, irrigation and lawn equipment, and service for those twenty-six million American homes that contract out their lawn care. But the pressure to conform is often more immediate. Neighbors who tolerate shaggy lawns are often thought of as laggards, negligent of their civic duty. As Michael Pollan, author of *The Omnivore's Dilemma*, notes: "That subtle yet unmistakable frontier, where the crew-cut lawn rubs up against a shaggy one, is a scar on the face of suburbia—an intolerable hint of trouble in paradise."

In some places the stigma of the shaggy lawn is fading, however. The Freedom Lawn may even garner some respect. Four blocks from my house Kathy Wilson proudly posts an official sign proclaiming her yard a Backyard Wildlife Sanctuary. A former owner of Kathy's property earned this distinction from the Washington Department of Fish and Wildlife. It required an assessment and map of the yard's resources—water, food, and shelter—for native wildlife. Kathy is thrilled that her yard is certified, and she continues to learn about her native plants and century-old tree stumps that she was encouraged to leave. Certification suggested, but did not require, other proconservation practices, such as mulching plants and harvesting rainwater, reducing or eliminating lawn, removing invasive plants, controlling cats, and blocking access to soffits and eaves by house sparrows and starlings. Our studies concur with all suggestions except the last one; we have no evidence that sparrows and starlings harm native birds, and rarely do they venture into nearby woodlands. But this issue matters little to Kathy, who is ecstatic about her small lawn, native plants, and the little care they require. Rather than buy bulbs and nursery stock to plant each year, she lets nature provide seedlings. She enjoys time and money that she would have spent on yard care with her son exploring their backyard nature.

I take a quick stroll around Kathy's yard. What I see—an abundance of native plants and little lawn—mirrors the findings from a larger assessment of backyard sanctuaries conducted by a group of University of Washington undergraduates. The towering big-leaf maple trees are free of invasive English ivy, a common scourge in these parts. There is just enough lawn for kids to kick around a soccer ball or play tetherball. The yard is far from pristine. Invasive groundcover spills out of the backyard flowerbeds, and there are nonnative plants with showy flowers and berries. But bees and hummingbirds, I learn, extensively use even these. To me, Kathy's yard is an example of how doing less can accomplish more. Conservation need not be onerous. In fact, natural processes do most of the yard work here, for free. The mixed flock of chickadees, nuthatches, and kinglets surrounding me, as well as the native Douglas squirrel that chatters from above, confirms that Kathy is a good neighbor.

Certification programs, such as the one offered by the State of Washington, are available for most U.S. residents. A quick search of the web reveals two national programs sponsored by the National Wildlife Federation and the Humane Society of the United States, as well as state-specific programs from California to South Carolina. Some cities even have their own programs—for example, Atlanta and Athens, Georgia, and Portland, Oregon. If you are interested, check with your local Audubon Society or state wildlife/natural resources agency for more information. Whether you certify or not, you will learn how to transform your yard of grass into a local wildlife haven.

Replacing part of your personal lawn with more diverse plantings improves the living space of birds, but architects, developers, relators, city planners, and planning commissions can do even more. Adopting "neighborhood conservation designs" preserves open spaces by clustering homes on nontraditional, small lots with little area for lawn while maintaining the housing density allowed by zoning laws. Landscape architect Randall Arendt, a conservation planner who leads the company Greener Prospects, details this approach in his practical and well-illustrated book *Conservation Design for Subdivisions* (1996). Akin to planned developments surrounding golf courses, conservation designs begin with developers working with regional planners, engineers, landscape architects, and wildlife professionals to identify approximately half the area that will remain open. Open space might be wetlands, waterfronts, important cultural areas, and steep slopes that are off limits to building in conventional neighborhoods, but also forests and fields of sufficient stature to sustain wildlife populations. Open space is recognized as a valuable amenity to residents' well-being and home value that might be amended with trails, playgrounds, and community gathering sites. In so doing, designed open space conserves wildlife, fosters community, and reduces the demand for private lawn space.

As I walk from my home to visit Kathy Wilson's backyard sanctuary, I pass through two neighborhoods built a decade ago. One is a traditional design, the other a conservation design. Each harbors a rich birdlife, exceeding that found in a nearby 150-acre forest. I see benefits for certain species in each design. Woodland species thrive in the conservation neighborhood, whereas a few species that forage on lawns flock to the traditional site. Although the number of types of birds has remained constant over the decade during which I surveyed the traditional neighborhood, diversity has steadily climbed in the conservation neighborhood. There, I actually encountered eight more species in 2010 than in 1998 when the site was entirely forest (an increase of 30 percent). What's more important is that nearly every sensitive forest species remains. Many species that require substantial prey or dead trees, such as Cooper's hawks and pileated woodpeckers, nest successfully. Such is not the case in the traditional neighborhood, where large lawns have replaced most of the former forest. A 2008 study suggests that the residents of the conservation-design neighborhood are also more satisfied, active, and rewarded with steadily rising home values than are those in the traditional neighborhood.

Encouraging the vegetation that birds and other wildlife require is a huge, but first, step in enabling subirdia to retain its riches. By increasing the diversity of plants in our neighborhoods, we can all increase the variety and health of wildlife that tolerates our presence. Providing shrubs and trees, especially native species that offer the nectar, nuts, berries, insects, and fruits that native animals desire, increases food resources. Thickets, brush piles, rocks, standing dead trees, and logs contribute cover from predators and safe nest and roost sites. With these basic elements present, we can now concern ourselves with increasing the ability of our urban spaces to support larger populations of birds and other animals. To do this requires us to increase the survival and reproduction of these species, which is also contingent on providing a cat-free place to live.

Keep your cat indoors! Our favorite felines are called "house cats" for a reason. Inside a house, with good veterinary care, a cat's life is free from parasites, disease, predators, injury, and deadly accidents that torment free-ranging mousers. Living inside extends the average cat's life by a decade. Keeping your cat inside and resisting the urge to feed strays or support programs that return spayed and neutered feral cats to free-roaming colonies also spares billions of wild animals needless death, disease, and harassment. A century ago, the Massachusetts state ornithologist, Dr. Edward Forbush, recognized cats as a major limiting factor to urban birds. Now we know that a free-ranging cat kills one in ten wild birds. Birds may be disrupted during nesting by cats and other domestic species, notably dogs that run free in parks, but disturbances are rarely lethal. In contrast, some birds that avoid the claw of the cat suffer increased nest failure and are exposed to lethal pathogens, such as the one that causes toxoplasmosis.

Those who wish to improve their yards for wildlife by keeping their cats inside may be interested in the American Bird Conservancy's "Cats Indoors" program. The program advises cat owners on how to keep their indoor cats happy. Starting with a kitten raised inside is the best strategy, but even older cats adapting to a life inside will appreciate a window shelf from which to watch birds, daily play with their owners, kitty grass for grazing, and access to an outside enclosure, such as a screened porch. Regardless of where you live, you can help wildlife by not allowing your cat to roam outdoors and by helping those you know who support a free-roaming cat to better understand the perils of doing so to tabby and nature.

Make your windows more visible to birds that fly near them. After death by cat, death by collision is the second most deadly and preventable force affecting urban birds. Though each residence in the United States is estimated to kill only a few birds each year, the sheer number of residences means that a quarter of a billion birds likely die each year by smashing into our homes' windows. Another ten million birds suffer the same fate in Canada. Some birds, such as ruby-throated hummingbirds that zip in and out of feeders we often place near windows, are disproportionately affected. Researchers (such as Dr. Daniel Klem Jr.) who pick up dead birds around windows for a living are helping to develop bird-friendly glass. Promising discoveries include coatings and imbedded patterns that reflect ultraviolet light that birds, but not people, can see. These reflections would warn birds of an obstacle without obscuring our view. The economics of supply and demand and the knowledge that some birds don't look where they are flying have delayed widespread availability of ultraviolet-reflecting windows. Beautiful decals in a variety of shapes and sizes that reflect ultraviolet light are available from specialty bird-feeding stores, but other less high-tech strategies also exist.

Applying artistic etching, decorative frostings, or simple striping to large windows is particularly effective. Obscuring as little as 5 percent of a window can significantly reduce bird strikes. "Bird tape" is available for striping from the American Bird Conservancy, but ordinary masking tape or even soap works fine. Vertical or horizontal stripes spaced from four to six inches apart, each of which need only be one-eighth of an inch wide, are most effective.

American robin killed after colliding with a window

Decorative grids that effectively break a large sheet of glass into smaller four-by-six-inch panes would also be effective. Outside screens and shades may be practical in some situations, and certainly are visible to birds. Lighting a window to reduce reflection during the day and providing shrubs outside windows rather than luring birds to houseplants inside windows are also good practices. The typical hawk and falcon decals that many windows bear are not effective, however. The key to preventing collisions is to make the window visible, not scary.

Keeping birds from battering into large commercial buildings, many of which feature immense glass facades, is a challenge in many cities. Increasingly, buildings in bird-rich areas are being required to incorporate bird-friendly designs, especially in terms of their windows. In Canada, a 2013 court decision rendered bird-killing windows a breach of national environmental law. In Chicago, San Francisco, Oakland, and throughout Minnesota, buildings that seek federal funds or environmental certification are required to reduce the collision risk that their windows pose. In response, architects are incorporating into their designs sunshades, less reflective windows, and windows with texture or opaque portions to increase visibility to birds. Reduced night lighting, which is especially attractive to migrating birds, often luring entire flocks into windows, is another part of a bird-friendly building design that deserves its own commandment.

Do not light the night sky. Light offers security, hope, and joy to people. We feel safe in a lighted space. We enjoy the colorful lights of a city from a prominent overlook and take a certain amount of pride in the industry that has enabled the outpouring of energy. Our sprits rise at the sight of holiday lights. Light's magic, however, has a dark side. Today, cities glow eight to nine times brighter than natural landscapes. Artificial light threatens to erase the night from much of the developed world. This phenomenon has astronomers on the run, seeking darkness where they can study the night sky. They take increasingly long trips because traditional observatories, such as Mount Wilson in the mountains above Los Angeles, are awash in city light. Artificial night light is also increasingly recognized as disruptive to all forms of wildlife in our cities.

Like moths drawn to a flame, birds, especially those that migrate at night, can be attracted to the light of buildings. Many of them die from colliding with towers, wires, windows, and walls. Collisions are especially frequent on cloudy nights when birds cannot use the movements of stars as navigational aids. Under these conditions birds rely on detecting the earth's magnetic field, in part using vision. This sense is compromised by the presence of red or white lights, which often adorn a city's highest towers and antennas. Other birds, reluctant to leave the lighted area, circle until exhausted. Some die in the predatory grip of night specialists, such as owls. Hundreds of dead and injured birds can be found in major cities every morning at the base of lighted skyscrapers. Experts estimate that nearly seven hundred thousand die annually from skyscraper collisions each year in the United States and Canada alone. Upwards of seven million may die each year in the United States and Canada when they strike towers. Warblers, especially the black-and-white, worm-eating, golden-winged, Canada, and Kentucky varieties, are frequent casualties because they migrate at night.

Some city light is needed for public safety and efficient commerce, but much is emitted needlessly in ostentatious displays that erode the quality of the habitats created in our yards, parks, and recreational spaces. In terms of the number of birds killed, excessive lighting can undo our best efforts to diversify land cover, keep cats indoors, and increase the visibility of our windows. Reducing the harmful effects of luxury lighting is something that everyone can do to improve our built environment's ability to support birds and other animals, whether you live in a small flat or are entrusted to plan the growth of a major city.

Three simple concepts can greatly reduce the harmful effects of light. First, use light only when necessary and then from as dim a source as possible. Bright bluish light is the most disruptive, and softer yellowish light is least disruptive. Second, where warning lights are necessary, the use of blinking rather than steady emissions and blue or green rather than red or white lights reduces collisions. Regulatory agencies in the United States recognize the benefit of blinking lights and allow many tower operators to turn off steady red lights and replace them with fewer, blinking ones. This measure costs little and pays operators huge dividends by reducing power bills in addition to saving birds. Third, face outside lighting downward, not skyward, and especially avoid illumination that shines horizontally. Upward and horizontal lighting illuminates vast areas, often including unintended spaces that could easily be kept dark by a more focused beam. Batman might be clueless about crime in Gotham City without its Bat-Signal searchlight, but the real bats of Gotham and the birds migrating above would be safer in the dark.

Do the streetlights where you drive dim or turn off late at night when traffic is slack? Are your stadiums and sports fields brightly lit even when nobody is playing? Are skyscrapers and monuments aglow all night in your city? Reducing these major sources of light pollution, and energy waste, can vastly improve the night environment. Technology and willpower are making a difference. Lights Out Indy, for example, is a citizen-driven initiative that is helping Indianapolis, Indiana, save birds and energy by going dark between midnight and dawn during bird migration seasons. Building owners and tenants work together to reduce nonessential lighting of their buildings. Similar programs exist across the United States and Canada, though not in other countries. Rigorous evaluations of effectiveness are not available, but anecdotal reports suggest that darkening cityscapes during migration really pays off. Bird collisions at one Chicago building dropped by 80 percent after the lights were turned down. And with concerned residents patrolling the streets for the wounded, many birds' lives are saved. Even the casualties are put to good use, ending up in scientific repositories. The citizens of the grassroots group Chicago Bird Collision Monitors, for example, sent nearly nineteen thousand specimens to the Field Museum between 2003 and 2010.

Take a night walk around your home with an eye toward finding places you can darken by removing, repositioning, automating, or dimming your existing lights. Try working with your neighbors, landlords, and city facilities managers to darken larger areas. And if you want to join a Lights Out program in your city, check the web to see what is already going on and learn how to participate, or contact your local Audubon Society. If you're interested in starting your own program, the National Audubon Society has an online tool-kit that will help get you started.

Provide food and nest boxes. For all the challenges that cities pose to birds, they also offer bountiful riches, particularly in the form of a million tons of birdseed. This annual subsidy from U.S. residents alone, plus water and shelter available in cities, increases bird abundance by bolstering overwinter survival and reproduction. Birds such as black-capped chickadees, which are common denizens of subirdia, derive 20 percent of their daily energy requirements from bird feeders if they are available. This fraction, while substantial, suggests that birds using feeders do not abandon other natural foods and do not become overly reliant on feeders. Indeed, when feeders were removed from chickadees, their survival dropped, but only back to normal levels. Such tangible benefits of bird feeding compel me to try keeping my feeders well stocked, but also put me at ease when they are empty.

Bird feeding benefits not only birds but also people, by bringing them closer to nature and by pumping billions of dollars into local businesses each year. It provides one very important connection between the people and the wildlife of subirdia that forges our mutual destiny as coevolving partners. But before you rush out and get a feeder, let's quickly review some best practices.

Most birds that use feeders prefer black oil sunflower and white proso millet seed. So a balanced mix of these is a surefire way to begin attracting birds to your feeder. Adding other types of feed to your yard—nectar for hummingbirds, thistle for finches and siskins, peanuts for nuthatches and tits, whole popcorn kernels for pigeons and doves, and suet for woodpeckers— will increase the diversity of birds you can attract. Most birds readily take to open-platform feeder designs, but some such as goldfinches require special-ized tube feeders that favor their needle-sharp beaks. Placing feeders near cover and away from windows increases their safety and therefore their attrac-tiveness to birds.

Keeping your feeder full is only part of effectively supplementing the diets of wild birds. Feeders need to be kept clean and dry, and food needs to remain fresh to reduce exposing birds to pathogens, such as salmonella, and communicable diseases, such as conjunctivitis. Your local Audubon Society or specialty feed store can help you decide what type of feeder and food work best.

The British Trust for Ornithology recommends feeding birds year-round. Consistent food improves not only survival but also breeding success, because overwintering birds with access to food are kept in top condition. It makes sense to me to let feeders go empty every once in a while, however; uncer-tainty in food supply is not uncommon in nature. When the feeders are empty in my yard, birds turn to berries, nuts, and insects in my bushes and trees. Filling feeders for a single day's use and letting them remain empty at night is a good strategy to reduce rats, which otherwise help themselves to bird food each evening. If you live in bear country, extra care is required to keep from subsidizing backyard bruins. In this case, feeding birds only during winter when bears hibernate is advisable.

Offering food from feeders or "tables" is not the only way you can supple-ment urban birds. Some birds, especially crows, jays, magpies, and nutcrack-ers, quickly learn to recognize a person who reliably feeds them. They will eagerly mooch nuts from a windowsill or deck. Whole nuts, unsalted in the shell, work well for these birds, which cache surplus food for later use. Just remember that these birds are smart, and they will tap your windows, ring your doorbell, and follow you hither and yon for food once they figure out your routine. They have been known to upset less-intrigued spouses and neighbors with their persistence. But if you want to forge a unique association with a wild bird, this is your chance. My colleague Professor Marc Miller, an anthropologist, helped me study this phenomenon in Seattle. We discovered that crows and the people who feed them share a language of sorts, reading and anticipating the actions of one another so that both parties can form strong, mutually reinforcing bonds.

In urban settings, where dead trees are rare, birds also benefit from the provision of nest boxes. Selecting a box of appropriate size for native, secondary cavity nesters, placing it in a suitable location, and maintaining it can teach a lot about the unique requirements of swallows, chickadees, bluebirds, wrens, and others. Doing these things provides nest and roost sites for birds even in the most urban settings, such as on an apartment wall, highway sign, or downtown high-rise. As with feeders, the safety of bird boxes is improved by nearby cover. Birds that nest on ledges can also be enticed to breed on small shelves or other flat surfaces that are best provided under eaves. Barn swallows and thrushes, for example, readily take to such structures.

Nest boxes are good, but nothing beats a real, honest snag when it comes to keeping nest and roost sites in the urban environment. Preserving dead trees, or even dead limbs on otherwise healthy trees, is the best way to ensure a steady supply of cavities for the species that require them. You can help the natural processes that create snags as well by killing trees where you have the space to safely do so.

In my overstocked acre of forest, I've created twenty-five snags by topping selected trees. Doing this provided sunlight to my forest floor, firewood to help heat the house, and some great woodpecker trees. The snags I made ranged from ten inches to nearly three feet in diameter and towered up to seventy feet. They provided perch sites immediately and a long-lasting supply of potential cavities after decay set in. I worked with a friend who won the 2005 International Tree Climbing Championship, Dan Kraus, to fashion snags especially attractive to wildlife. Dan cut slits in the snags that bats might use for roosting, started holes to speed up the decay process, scarred up some of the bark as lightning might, and trimmed the tops to look like natural breaks. Red-breasted sapsuckers, hairy woodpeckers, northern flickers, and pileated woodpeckers all visited the snags within three months of their creation. These species hammered out beetle larvae and termites from beneath the bark, but none bred. Finally, in year three, hairy woodpeckers broke the ice. Fungi and the workings of woodpeckers softened up the snags enough to allow chestnut-backed chickadees to nest in two snags in year four. At this time the bark also started to peel away from a few snags, and a brown creeper placed its nest securely between the tree bole and bark flap. I'm still waiting for pileated woodpeckers to dig a house in my larger snags, but they require a pithy core and solid wooden wall that is possible only with heart-rot fungi. This process will take more time. Until the pileateds drill, I enjoy the occasional visits by red-tailed hawks, great horned owls, and bald eagles to the large snags. It is a long process from dead tree to woodpecker home, but I encourage you to make a snag. And then watch the steady progression of fungi, insects, and woodpeckers turn it into a real bird condominium!

Birdbaths and other water features are essential supplements for urban birds in dry environments, but even in the temperate rainforest where I live, they are attractive. Cover is especially important around water sources, because bathing birds are vulnerable to predators. Sometimes it's not the water, but what's in it that birds seek. Great blue herons regularly help themselves to the goldfish in my neighbors' fishponds! You can protect your fish by giving them underwater cover.

Supplementing birds with food, water, and shelter might seem unnatural, but to me this is an essential way that we can foster resilient populations of urban wildlife. The willingness of people to provide, and animals to receive, is a signature of the urban ecosystem. By providing subsidies, we enable some species to enlarge their populations and thereby better absorb the inevitable losses they will suffer to cats, windows, cars, and pollution in all their many guises. Surplus birds may not increase breeding density in the best habitats, which in territorial species are likely to be fully defended by existing pairs. Backups, or "floaters" as they are often called, provide important buffers against random fluctuations in population size. Abundance is also attractive. Professor Amanda Rodewald proposes that the presence of common species is attractive to a variety of rare suburban birds. The common birds are reliable indicators of a habitat's quality. In this way maintaining large numbers of urban adapters and exploiters can help attract some avoiders. This not only increases the conservation value of subirdia, but also the complexity and adaptability of our shared ecological web. A diverse web of life that is able to evolve is the true sign of a healthy, if somewhat artificial, ecosystem. Larger populations of prey, for example, support food webs that feature many native predators. Living with these predators, however, can challenge our moral fiber.

Do not kill native predators. My wife, Colleen, is a biologist, so when she called and excitedly told me there was an unusual large, white bird sitting atop the cell tower a mile from our house, I grabbed my scope and hit the road. As I got the stout animal in focus, its bright yellow eyes met mine. A snowy owl! What a treat it was to finally see this visitor from the north, especially so close to home. The bird had long been on my shame list; despite looking for it many times, this was the first one I had ever seen in the flesh. I thanked the arctic for producing a poor annual crop of voles, the owl's typical prey, which was the reason this bird and many other snowy owls invaded the United States in 2005. Similar invasions occur every four years throughout the northern hemisphere, with owls showing up in Asia, Europe, North America, and even on ships at sea.

I cannot imagine anyone who would not marvel at a gorgeous white owl that wandered into his or her neighborhood. Indeed, our local papers were full of reports about the owls and the birding fever they had ignited. Coexisting with other predators is dicier.

Sugar and Marcia were fixtures in the neighborhood. Marcia rode her scooter as Sugar, a plump rock star of a Jack Russell terrier, strained at the end of the leash to snuffle in the ditch or nap in a sunbreak. Occasionally, Sugar would scoot out of the house and follow her terrier instinct into the wooded ravine in Marcia's backyard. Such a breach of security nearly cost Sugar her life in 2010, however. The wails coming from the wooded patch were not the usual baying of a dog on an adventure. These were clearly the sounds of an animal in distress. A neighbor, sensing trouble, got his gun, bolted toward the screams, and shot the coyote that had pinned Sugar to the leafy soil. Sugar somehow survived. The coyote did not.

I was glad for Sugar and Marcia, but I lamented the loss of the coyote. I loved seeing and hearing coyotes in the neighborhood, and I felt they were essential to the health of our ecosystem. Where coyotes roam, cats are rare and bird populations are healthy. Conserving carnivores is challenging everywhere, because while they inspire some people, they horrify others. Both perspectives are valid and not easy to resolve, especially where we live or raise livestock. But subirdia needs the natural checks and balances that predators provide, as they cull the diseased or weak and reduce the overly abundant. Let's consider a few ways we can limit the risks, however minor they may be, that wild predators pose.

If you share the landscape with cougars, bears, wolves, coyotes, dingoes, and other large carnivores, you can reduce their interactions with your pets and livestock with good animal husbandry. Dispose of garbage, compost, and offal in such a way as to not attract predators, most of which are also scavengers, to the places where you keep domestic animals. Bird feeders and pet foods can also lure some predators into your yard, so be careful when and where you provide feed. Your small pets will be safe inside at night, and you can increase the safety of your larger animals by providing outside shelter. Enclosing small pastures, barns, or cages with electric fencing is highly effective at deterring even the largest predators.

As we learn to live with potentially dangerous predators, we should also take care to limit the dangers we pose to these thrilling creatures. One of the easiest ways you can limit your effect on hawks, owls, and mammalian predators is to exercise restraint in the application of pesticides. The toxins we use to control insects and rodents are deadly to small birds and mammals. Today's potent, second-generation anticoagulants, such as brodifacoum and bromadiolone, that are available to homeowners for control of mice and rats have killed owls and a wide range of wild and domestic mammals that either ingest the substances directly or ingest sick rodents. Survivors may also be more susceptible to disease, such as mange, and therefore increasingly likely to be involved in dangerous encounters with people.

I've spent considerable time battling the rats that try to sneak into my house and shed. I know their late-night scampering can easily lead one to the rodenticide isle of the hardware store. But before you give in, make sure you have found and sealed all the places where rodents can enter your home—even the tiny ones through which you think a rat could never squeeze. Then, try to solve your problem with snap traps. If you must use rodenticides, follow the package directions carefully and completely, or call on a professional exterminator. Regardless of how careful you are with toxins, some are certain to find their way into nature and harm the rarest and often most endearing residents of subirdia.

As we foster our own healthy urban ecosystem, we should also consider its unique qualities and strive to maintain the individual character of our yards, neighborhoods, and cities. This requires a wider view of the urban ecosystem.

Foster a diversity of habitats and natural variability within landscapes. Within our gardens, parks, and other small green spaces, birds benefit when nonnative turflands are diversified to include native ground, shrub, and tree cover. Over larger expanses—between neighborhoods, cities, regions, and continents—birds also benefit from variety, not uniformity. As the spatial arrangements of built and open lands become similar across cities and suburbs of the world, we favor the same, limited set of adaptable species, such as the fab five, everywhere. A greater variety of birds would enrich our lives if developers and planners sought variety instead of dividing the land into repetitive and regular patterns defined by an oft-repeated grid or spiral of roads. Interspersing neighborhoods of dense, evenly spaced homes with those featuring clustered lots and substantial open space increases a city's diversity. In Seattle, for example, Bewick's wrens, bushtits, and black-capped chickadees thrive in dense, older developments with decorative shrubs and many deciduous trees. Pacific wrens and chestnut-backed chickadees occupy the conifer buffers featured in newer, outlying clustered developments, while Swainson's thrushes and other edge specialists lurk in the boundaries of extensive salmonberry fronting small woodlots and parks. Beyond subirdia, others respond to more extreme conditions. The sparse shrubbery of strip malls provides a place for Brewer's blackbirds, while crows and gulls patrol the paved urban core. Simply not doing the same thing everywhere increases the diversity of habitats and therefore the variety of birds that exist within cities.

As we seek to reduce uniformity within a city, we must also consider how to reduce it between cities, especially those in distinct ecological regions. Consistent and unimaginative landscaping practices are a major homogenizing force that increases the chances that many cities will harbor the same few species of birds. At regional and continental levels, retaining native vegetation and landscaping to conform with, rather than contradict, this natural background can stall the march to uniformity. To date, homogenization is most evident in the frequently disturbed, highly modified parts of the city—suburban lawns, bare land, and paved downtowns. Scruffing up these places with native grasses, shrubs, cacti, and trees not only saves energy and natural resources by reducing maintenance, but also helps keep habitat locally diverse and regionally distinct. Many native plants could easily be retained if developers and builders carved lots carefully from existing vegetation rather than pushing it aside to ease their operations. In this way, neighborhoods in Phoenix, Arizona, for example, could retain iconic saguaro cacti and palo verde trees to distinguish themselves from neighborhoods among the coastal chaparral of San Diego, California. By scraping entire lots bare and then replanting them with a standard mix of nursery stock, uniformity among neighborhoods and cities rises and bird diversity declines.

General practices that reduce the uniformity of cities will benefit a wide variety of birds, but special features of disproportionate importance sometimes require individual attention as well. Monroe, Washington, for instance is home to around twenty thousand migrating Vaux's swifts each spring and autumn, because it retained a large, brick chimney once used to vent heating exhaust from a school's furnaces. The swifts, no longer able to find large, hollow trees to roost within, instead swirl en masse into the old chimney. The

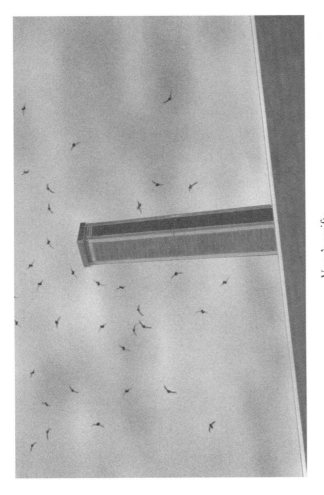

Vaux's swifts

community's shoring up of the teetering structure, rather than demolishing it, has provided swifts a critical resource and allowed students and local residents to enjoy a yearly spectacle. Similarly, by not filling and developing a flooded cottonwood grove, Kenmore, Washington, has provided a unique nesting area for its urban great blue herons. The elegant arrivals and departures of the massive birds thrill commuters waiting in the adjacent bus lot. Small buffers around sea cliffs enable other colonial birds to fill the skies of many coastal cities. Northern fulmars, moderate-sized relatives of albatrosses, for example, can be viewed at close range as they tend their eggs and young on the grassy cliffs of St. Andrews, Scotland. Even ruins, saved for their cultural importance, help distinguish a city's biota. More than 40 percent of all lichens that exist in England can be found on the old stone walls within London. Retaining unique urban legacies trims the urban tree of life with especially bright ornaments.

Keeping neighborhoods and cities distinctive by emphasizing their natural features reduces biotic homogenization, but other design principles aimed at reducing impediments to animal movement in particular are especially important to those that, unlike birds, cannot fly.

Create safe passage across roads and highways. Roads increasingly crisscross Earth. In 2013, the U.S. Central Intelligence Agency measured nearly forty million miles of roads, both paved and unpaved, on the planet. If we could aim these tracks toward our moon, they would stretch there and back more than eighty-three times! Though most birds safely navigate this tangle of concrete, gravel, and dirt each day, the world's reptiles, amphibians, and mammals fare much worse.

A slight wrinkle in the road caught my attention as I drove on Junipero Serra Boulevard along the Stanford University campus in Palo Alto, California. A bright yellow and black "newt crossing" sign told me this was no ordinary speed bump. Rather, it concealed a culvert that was one of three amphibian tunnels under this road installed by the university to allow California tiger salamanders, Pacific treefrogs, and western toads to migrate safely between their upland winter homes and the wetlands they require for breeding. It was a more technical solution than the road closed in New Jersey on behalf of spotted salamanders, but both strategies are simple, elegant, and effective at enabling many creatures to safely cross our extensive network of roads.

Driving at slow speeds and creating small or temporary crossing lanes probably spare the lives of millions of herps and mammals in suburbia, but more extreme solutions are needed along the high-speed freeways that connect our cities. "Moose-proof fences" funnel moose and other large creatures, such as bears, deer, and wildcats, to gaping tunnels where they can cross beneath busy northern highways. This increases safety for the animals and the drivers, and saves money. In other places, such as Banff National Park in Alberta, Canada, scenic bridges are built exclusively for wildlife passage.

Road right-of-ways often include substantial undeveloped spaces that can also benefit wildlife. Enabling natural grasslands and shrublands to fill in along the road network's verges and medians may benefit sensitive prairie and chaparral species. Taking care to limit disturbance of these areas during nesting is another simple way roads can be made less deadly to birds. In Canada, for example, twenty-five thousand eggs and nestlings are likely destroyed by roadway mowing each year.

One day we may move about cities in suspended gondolas or through sealed pneumatic tubes that minimally affect wildlife. Until that time, designing, retrofitting, and maintaining roads that consider the needs of the animals we live with could go a long way toward increasing subirdia's biological diversity. Converting some roadways to pedestrian walkways and providing naturalistic trails within cities would increase their livability for humans and other creatures. Ecological road design, however, is part of a larger connectivity issue that is best solved with coordinated regional planning.

Ensure there are functional connections between land and water. Connectivity among the green and blue spaces within a city—its outdoor recreation sites, parks, wild tracts, rivers, lakes, and wetlands—while beneficial to birds, is absolutely essential to the well-being of fishes, reptiles, amphibians, mammals, and many of the insects upon which they and we depend. These homes to our furry, scaly, and slippery neighbors also add distinctive character and charm to cities, forming what Frederick Law Olmstead called the Emerald Necklace in Boston, for instance—a park system that includes gardens, fens, ponds, and waterways. Such things are literally the ecological lungs of the city that provide stress relief, natural resources, clean water, pleasant climate, and noise abatement to urban people. Unfortunately, many cities do not have an Emerald Necklace in their jewelry box, or do not formally recognize the one they have. Working with planning authorities is an important way that builders, developers, and concerned citizens can provide what much of nature requires.

In 2008, Harvard professor Richard Forman assessed the "emerald networks," as he called them, of thirty-eight regions that included representative cities around the world. Only one-quarter of these regions, those surrounding cities such as Barcelona, Spain; Canberra, Australia; Iquitos, Peru; and Stockholm, Sweden, had a fully interconnected emerald network, which Forman interpreted as a set of large natural patches that effectively connected at least one hundred square kilometers (sixty-two square miles). Three urban regions, those that included London, Chicago, and Bucharest, were entirely devoid of natural landscapes. But the majority of regions had several, often including substantial wooded areas near the city center and many small, linear forests elsewhere. Gaps between upland natural areas were common; in one-third of the regions at least one natural land parcel was isolated by a single major gap, often a highway. The lack of wetlands within urban regions was prominent. Major wetlands were absent from urban regions with cities of more than eight million people and generally scarce in regions hosting cities of two to eight million. Urban regions around smaller cities, those with fewer than half a million people, typically supported numerous wetlands.

Beyond assessment, Forman offered a set of widely applicable strategies to create and maintain a region's emerald network. These provide starting points for planners and policymakers interested in increasing the functionality of their urban regions for animals that must walk or crawl across roads and between aquatic and terrestrial domains. Regional planning by a confederation, coalition, or agency that results in a binding, spatially explicit master plan (often called a "comprehensive plan") is needed to articulate the long-term vision and desired future conditions of an urban area. For the betterment of wildlife and people, this plan begins with mapping existing jewels in the network and then devising protection, restoration, or connection plans for each with an eye toward defining a fully interconnected network.

Reserves, parks, municipal watersheds, small urban farms, lakes, rivers, and wetlands constitute the backbone of the network. In these, native plants either occur or can be easily restored. Including a large central park or an impressive park on the edge of the city that protects cultural and natural history can enhance the economic value of the collection. Once the backbone is revealed, connections that allow walkers and wildlife to move between the emeralds are developed. These might include buffers along waterways, vegetated corridors, golf courses, or a series of small parks strung together by a trail that navigates an abandoned transportation route. Popular "rails-to-trails" programs in the United States, Europe, and Australia provide the latter sort of connection. Where existing roadways block people and wildlife from moving between emeralds, underpasses or overpasses, such as those described above, need to be devised. Restoration of water quality and riparian vegetation along prominent rivers that connect emeralds increases their attraction to tourists and recreationists, which benefits local economies. Preserving floodplains and restoring the usual network of small wetlands that occur in them can enhance the importance of the emerald network to aquatic life. With a natural emerald network mapped out, places to encourage and limit growth are then important to identify.

Effective regional planning faces many challenges, though it is far from impossible to accomplish it. Thirteen of the fifty U.S. states, for example, mandated some form of urban growth management. Many other local communities also have a strong record in visioning and planning their future growth. Boulder, Colorado; Oslo, Norway; and Barcelona, Spain, offer good examples that might be put into practice for the animals that live within your city. These places are moving toward what Professor Tim Beatley calls "biophilic cities," where nature is explicitly considered in the design of buildings and in the arrangement of built and more natural space. The result improves the health of people and their ecosystem.

Like other ethical codes humans adopt, living a life that birds and other wildlife might appreciate can be expressed as a "golden rule":

Do unto your land, and the natural web of life it sustains, as you would have the land do unto you.

Or more simply,

Love thy wild neighbor.

As Steve and Suzie Humphrey restore their cow pasture to cloud forest, they experience the land's response. Almost immediately, a bounty of flowering plants cushions their view and attracts gorgeous birds and a symphony of song to their doorstep. But the Humphreys are fully aware that they will not live to experience the land's most grateful response. Earth's ability to absorb our excessive emissions of carbon dioxide is increased with each tree restored. This restoration in turn may eventually slow the rapidly changing climate that scientific consensus recognizes is seriously imperiling our coastal cities, frozen poles, and diverse web of life. It is this neighborly act that truly motivates Steve and Suzie's obsession. Living by ecology's golden rule isn't easy, and the benefits of doing so may not be felt for decades. But because of the millions of good neighbors who foster diverse environments, some of us enjoy cities full of wonder, and our descendants are more likely to inherit from us a planet of which we can be proud.

TEN

Nature's Tenth Commandment

If the connections between city and forest become recognized and nurtured, if we see cities and woods as the interdependent, uninterrupted landscape that the bears and coyotes know them to be, then we can find ways to protect both.

—Ellen Stroud, *Nature Next Door* (2012)

Slogging through the sand and mud on a day-long trek into the heart of Costa Rica's wilderness allows me time to appreciate the importance of large conservation reserves. As my students and I walked the long but unspoiled footpath into Corcovado National Park, we were filled with awe and anticipation. The weight of our packs seemed to lessen as every bend of the trail held the promise of a fierce predator, lumbering prey, or beautifully plumed bird. Without reserves isolated from human enterprise, these sorts of dangerous, tasty, and valuable animals are quickly extinguished. Indeed, barely two hours from road's end, we encountered a Baird's tapir. The three-toed horse relative used its pint-sized, flexible trunk to sniff out a tender *Psychotria* leaf (a relative of the coffee plant) from the bountiful salad growing in the tropical rainforest's understory. As we marveled at the tapir's odd form, it was wholly

Facing page: Baird's tapir

unimpressed by us. In our presence, it simply grazed in peace. Northward, beyond the reserve, where we had spent the past two weeks, no tapirs exist. Even the expectation of seeing one has vanished. The tapir, a classic avoider, absolutely requires reserves. Some of us humans also need reserves, without which our muscles soften, our minds dull, and our souls lose a bit of hope.

Identifying, designing, and protecting reserves are basic practices in the craft of conservation science. The optimal number, size, shape, connection, composition, and location of reserves are debated and modeled; but finally, they are set aside through purchase, easement, or government action. In this way, somewhere between 10 percent and 15 percent of Earth's land has been reserved, though less than 6 percent is strictly protected for conservation purposes. Most reserves are in places where humans have little at stake—high mountains and lands too rocky to till. Some are in biologically diverse areas, such as the Amazon River of Brazil or the beaches and mountains of Costa Rica, a country that has reserved nearly a third of its landmass for nature—and the human economy that it supports. Reserves in productive shallow oceans are especially rare; only about 1.5 percent of coastal marine areas are protected.

Isolating much of Earth's biological diversity from our destructive ways through a system of reserves is a challenging possibility. Expanding, monitoring, and compensating those displaced by a global system of protected reserves that covers 15 percent of Earth's most diverse land would cost about $300 billion per year. That is a far cry from the $6 billion per year the world currently invests in protecting existing reserves, but given that there are now nearly eight billion of us and likely to be ten billion by 2050, the price seems affordable. For about forty dollars a head, we could safeguard the most diverse lands on Earth. This fee would require substantial and creative collaborative public and private investment, most notably a significant flow of funding from the wealthy north to the biologically diverse south. That many of the world's most biologically diverse "hotspots" also include large urban populations poses challenges and opportunities. Even if today's people were willing to make this investment, which I agree is necessary, it is only part of the change needed to conserve our biological diversity.

But reserves, while necessary, are alone insufficient for conservation. Being place-bound, today's reserves may not be in the best locations given tomorrow's conditions. The iconic spotted owl knows this lesson only too well. In the 1990s, a team appointed by then-President Bill Clinton designed a series of reserves to protect the owl's old-growth timber habitat in the Northwest. Unfortunately, these reserves are failing today, because the spotted owl's larger and more aggressive cousin, the barred owl, has found them to its liking. A native to eastern North America, the barred owl has invaded the West, in part because of the trees that settlers planted in the northern Great Plains. Competition with barred owls was not anticipated when the old forest reserves were designated, so they have been unable to provide the safe refuge for the spotted owl that policymakers expected. Limitations such as this may be especially important as climate and settlement patterns change.

Reservation also does not guarantee protection for all. Most reserves are too small and too widely separated to sustain their riches. Today, many are still overly exploited for wealth or food, a trend that is expected to grow with increasing human populations. Where reserves protect the species that do not thrive in our presence, they may not protect the vast diversity of adapters, those animals that seek frequently disturbed, yet lightly settled, edgy lands that fringe our cities. For these reasons, conservation strategies built solely on reserves are inadequate.

There is another, more fundamental reason that reserves will fail to conserve biological diversity; most people will never experience their grandeur. By their very design, reserves seek to separate people from nature. Conservation, however, is a human enterprise that depends on the value humans put on nature. When humans live in the city and nature lives in distant reserves, nature becomes less relevant and more fearsome to people. This process has

Children feeding rock pigeons in Costa Rica

been called "environmental amnesia"—literally, we forget about our natural world. As we forget, we also devalue nature and lose a will to conserve it. At best, we come to accept lowered environmental quality as the new norm—shifting baselines by which we gauge further change. To remember what biodiversity is and why it is important, we must conserve nature close to where we live and work as well as develop distant reserves. The apparent contradiction that global conservation of biological diversity may hinge on an appreciation for nature in our built world has been dubbed "the Pigeon Paradox"—that is to say, we might have to learn to love "rats with wings" and other city dwellers to help us develop a conservation ethic that extends beyond, but also includes, the city.

Including the urban ecosystem, and within it subirdia, in a broad conservation strategy can cure environmental amnesia. The city is the place where we can foster a love of nature because it is where we experience nature. There, among the streets and buildings, ecologically aware citizens bring local and scientific knowledge to bear on pressing social and environmental issues. Doing so not only helps the animals now, but also may be essential to those of the future that will require that *Homo urbanus* value nature at a distance, despite living in the city. Those wishing to meet the future with a more civic conservation ethic should be guided by a tenth commandment:

10. Enjoy and bond with nature where you live, work, and play!

As we work to connect humans and nature, we have an opportunity to right a substantial environmental injustice. The urban poor have long suffered the worst of the city's pollution, noise, and grime. Their homes are packed into the most degraded portions of the city, where vegetation is sparse and birds are rare. Beyond the slums, even neighborhoods of low-income, single-family residences are biologically impoverished. In contrast, nearby affluent neighborhoods are cloaked in ornate vegetation that supports a rich diversity of birds. The unjust distribution of biological wealth within a city is a global phenomenon, as Dr. Will Turner and his colleagues discovered in cities in the United States, Japan, Italy, and Germany. In all these places, a vast majority of all urban residents live in neighborhoods that harbor fewer bird species than are found on average throughout the city. Turner surmised that "most of Earth's human population lives in biological poverty" and worried what this global separation of humanity from nature might mean. I share this concern, but also see a path forward.

Rather than forsaking the degraded lands where most people live, those with a mind toward conservation could engage citizens and civic leaders to improve their ecological condition. Restoring native vegetation to marginal lands and along urban waterways would be especially significant. In regions where cities are shrinking in population (for example, in the Ruhr Valley of Germany, in middle England, and in the upper Midwest in the United States), substantial open space now exists that could be made into biological hotspots for all city residents. This is happening where shrinking cities are revitalizing their downtown core areas. In Leipzig, Germany, for example, where nearly 90 percent of manufacturing jobs were lost from 1970 to 2005, urban renewal has converted many abandoned fields into diverse neighborhood parks. Beyond restoring the urban commons, incentives could be used to encourage those in denuded neighborhoods who own yards, even tiny ones, to augment them with trees and shrubs. Even a single tree can increase a family's comfort and reduce the electric bill in warm climates by offering shade. Trees and shrubs also stabilize the soil, reduce runoff that leads to urban flooding, and clean pollution and dust from the air. Revitalizing the green lungs of the inner city would increase biological diversity exactly where it is most needed—where people live. There, residents can reap nature's emotional, psychological, cultural, and economic value—and in so doing rekindle a passion for their wild neighbors.

Community gardens have taken root on vacant lands in many of our largest cities. Greening these lands is good for ecology, but even better for sociology. Individuals and communities gather and work in gardens to produce favorite foods and conserve local customs that help build community identity, alleviate stress, and demonstrate resolve after challenge. After the terrorist attacks of September 11, 2001, New York's community gardens were seen as resilient, living memorials to the loss. In New Orleans after Hurricane Katrina residents of the Tremé neighborhood planted trees to remember a better time and prepare for future challenges. Getting involved and sustaining life by restoring urban lands promotes participants' emotional and psychological well-being.

By planting and caring for urban greenery, citizens gain basic ecological information and share local knowledge. For instance, by pulling up invasive

English ivy, community groups in Seattle learn how nonnative groundcovers compromise urban conifer forests. Their actions improve the ecological function of the trees. In New Orleans, as majestic southern live oaks are restored, citizens learn about the value of trees to wildlife and the sense of place they bring to southern cities.

A restored urban green is only the starting point in bonding humans with nature. Parks, gardens, and other public gathering places could be enhanced with bird feeders and nest boxes that concentrate birds where people can most enjoy them. These critical supplements are rare where people rent, rather than own, their homes, so adding them to rental areas could be especially productive. When I birded New York's Central Park, I was stunned to happen upon a garden of bird feeders. Deep in the park, past the bowling greens and the Lake, as I entered the woods of the Ramble, I heard the metallic *eenk* of woodpeckers, the cry of a jay, and the soft whistle of a titmouse. Following my ears, I crested a slight rise in the primitive trail, and before me were baskets of suet, tube feeders, and platforms loaded with birdseed. The ground seethed with sparrows and juncos. Hairy, downy, and red-bellied woodpeckers clamored on the suet. American goldfinches, tufted titmice, and northern cardinals probed tube feeders stocked with sunflower and millet. An immature Cooper's hawk hunted the station, as if to emphasize that "Yes!," the web of life fueled by human subsidy is strong and intact. People crowded into this tiny space and watched with delight. Maintaining a feeding station in the heart of the city allowed visitors and residents alike to come face to face with New York's biodiversity. As stress fades, complete strangers converse and learn from each other about the birds and about the city. I'd like to think that the homeless man I saw sharing his scavenged slice of pizza with the birds also felt the quality of his life enhanced, even if just for a moment.

A biologically diverse neighborhood has more traditional benefits as well that may motivate some people to improve the ecological condition of their property. The consensus of studies from Europe and the United States is that the presence of trees—in the yard or in nearby forested parks—increases the price of a home by 5 percent to 10 percent relative to comparable homes without trees. Researchers in Lubbock, Texas, were more direct in estimating the value of birds to home sales in 2008 and 2009. Attracting an additional "desirable" bird species to a neighborhood increased the sales price of homes by $32,000. (Desirable birds were those that were relatively uncommon in Lubbock and associated with increased trees and shrubs in yards.) In this way a neighborhood that provided habitat for American robins, blue jays, and mourning doves might enjoy home values averaging $32,000 more than those in a neighborhood able to attract only two desirable species, such as northern mockingbirds and western kingbirds. Improving the vegetative diversity in your yard, nearby park, or rental property makes ecological and economic sense.

Green neighborhoods may also help break the cycle of poverty, crime, and unhealthy lifestyle that challenge many urban people. In Philadelphia, for example, gun assaults and vandalism declined in neighborhoods that restored healthy greenery to vacant lots. Residents of such neighborhoods also reported being less stressed and exercising more than did those living in neighborhoods that did not revitalize the ecology of vacant spaces. The practice of ecological restoration and the tangible health benefits it provides motivate residents to remain engaged in civic projects and thereby improve even more of their environment. This positive feedback can reconnect people with nature and improve the social and ecological health of cities.

Residents of biologically diverse cities that are actively engaged in ecological projects are resilient. They become familiar with their community's history and local knowledge and create new experiences through scientific inquiry. They forge partnerships with nonprofit organizations and governmental agencies that may be called on in times of emergency. They practice self-organization and stewardship that enhances resilience to ecological and social change. In short, they adapt to the vagaries of nature. Their strong social fabric and the attitudes it encourages better enable recovery after disaster.

Beyond the social benefits, increasing the ecological function of Main Street may draw customers to a more aesthetically pleasing city. It is important for business leaders to appreciate the fact that economy benefits from a healthy ecology. Ignoring this connection can outrage would-be customers. For example, in the late 1990s, New Yorkers protested against their mayor's proposal to convert community gardens to commercial properties. In response, the state of New York bestowed park status on gardens that met certain standards. Rather than banking on short-term economic gains that sacrifice nature, city planners realized that creating a biologically diverse city could sustain long-term economic gains and produce a more motivated and fit workforce.

Connecting the city with its historic wilder ecosystems pays large dividends. Billions of dollars in filtration costs are saved when forested watersheds are retained around a city's water supply. Seattle, Salt Lake City, and New York City, for example, all reap these economic benefits. Expanding urban forests might help cities in other ways as well. In downtown Indianapolis, Indiana, for example, businesses spend about $100,000 per year to discourage roosting aggregations of pigeons, starlings, and American crows. Before their investment, which funds a team of government bird busters, the feces from the birds had to be cleaned from the sidewalks and buildings each day at an even greater cost. But displacing the crows has other, unintended consequences. Keeping crows out of the business district forces them to roost in a nearby low-income neighborhood. Investment in an urban forest made attractive to crows, perhaps by expanding the green fringe along the city's White River, could provide a longer-term, ecological solution that allows the birds? adaptable and social nature to be celebrated while they fertilize a more socially acceptable location. Continuing to push the problem off on the less fortunate is a form of environmental injustice that may build resentment that eventually costs more than it saves.

The smile on the young Costa Rican boy's face as he offered a bit of food to the variegated squirrels and rock pigeons that swarm Alajuela's Central Park demonstrated the important effect our bond with nature has on our emotions. Wonder and joy brought the kids, parents, and grandparents together to value their urban animals. Similarly, in the summer of 2009, a grandfather and his two grandchildren stopped my car so that a mother mallard and her brood of waddling ducklings could safely cross our neighborhood street. Interacting with the common creatures of subirdia allows the new generation to learn about empathy from the older.

As we interact with a wider range of urban nature, we experience other emotions that offer unique teaching opportunities to our loved ones. Difficult lessons gleaned from the extinction of species in degraded lands or loss of life to a window, cat, or car can nurture an ethic of stewardship so future losses are reduced. Encountering what may seem like unsavory behavior in our urban web can broaden our tolerance of and appreciation for different lifestyles. Our helping ducklings is as valid and important as understanding that the hawk and fox that eat them are also essential components of nature's web. If an ethic of appreciating predators could be nurtured in the city, it might be transferred to wilder lands where today we seem unable to value most animals that hunt in order to live.

Using our daily encounters with plants and animals in the urban ecosystem as a place to cultivate a biophilic ethic and gain ecological literacy will equip future generations to better utilize and steward their world. Interpreting and showcasing the marvels of the urban ecosystem for its human residents are important roles for conservation and nature organizations to embrace. Many already do. The Cornell University Laboratory of Ornithology encourages people to join its Celebrate Urban Birds project, while Birdlife Australia promotes urban birds with its Birds in Backyards program. Museums and environmental learning centers offer camps to get young people outside. Festivals celebrate the natural cycles of birds across the world. One can experience the Brazilian Bird Fair in the heart of São Paulo, a metro area of more than twenty million people. Or take in the spectacles of flamingo and common crane migrations across the cities of Spain. Closer to my home, I enjoyed a "swift night out" that was sponsored by my local Audubon Society. There, the people of Monroe, Washington, watched tens of thousands of Vaux's swifts settle into the old school chimney and learned about these interesting birds and the fascinating native predators that also track their annual movements.

Subirdia is a bridge that connects the more urbanized parts of the city to the wilder country beyond the metropolis. As such, it is vital to sustaining the biological diversity that can enrich the lives and pocketbooks of urban people. Yet development often proceeds haphazardly, with little regard for ecological health. Change is afoot, as some planners and architects incorporate ecological sustainability into the design of buildings, neighborhoods, and cities. Those striving to create "biophilic cities" understand that not considering the amount and arrangement of green spaces that connect urban people with nature is inefficient and dangerous. When natural land cover—measured across areas the size of neighborhoods, metropolitan areas, or counties—drops to less than one-third of its historical extent, its ability to sustain native biodiversity crumbles. The inability to support nature accelerates below this threshold as connections between remnants of native land are severed, invasions of nonnative species increase, erosion carries away fertile soil, and temperatures rise.

The 30 percent threshold represents an ecological tipping point for subirdia. This was evident in my studies of Seattle's birds, but theory suggests that this threshold will characterize changing landscapes wherever they exist, from the tropics to the tundra. As with many tipping points, detecting the impending change and enlisting civic concern are difficult because all seems good right up to the cliff. In Seattle's subirdia, bird diversity remains high as forest cover drops from 60 percent to 40 percent of its former extent. Diversity declines slowly as forest cover erodes to 30 percent, but then the decline accelerates with each additional, though slight, drop.

Social and ecological systems change after tipping points are crossed, often in ways that naturally increase their resilience to future change. In this way banks are restructured after economic meltdown. Building codes are rewritten after earthquakes. Security measures are reevaluated after attacks or abuses of trust. In the face of disaster, human systems frequently reorganize by increasing our connection to nature. Community gardening has increased in postconflict Bosnia; park projects blossomed in Berlin where a wall once isolated residents; and after the earthquake that leveled much of Port-au-Prince, Haiti, Martissant Park was built for those who needed a positive space within which to gather, grieve, and reconnect. These responses give me hope that the benefits of our ecological resolve might engender more proactive responses in the future. Our existing civic organizations can push their urban ecosystems back from the tipping point by increasing, improving, and restoring the green spaces that prove so valuable in our most trying times.

Climate change in response to increasing atmospheric carbon dioxide is challenging life as we know it. In the carbon-enriched air, not only are temperatures rising, but also seas are acidifying, diseases are spreading, and the timing of basic biological events, such as bird migrations, are getting out of synch with the natural products, such as fruits and insects, that they depend upon. Curbing our combustion of fossil fuels is essential if we are to minimize these impending disasters. As we cut back on the use of fossil carbon, we can also follow Steve Humphrey's lead and convert some of the turflands of subirdia into native forests, or where forests do not naturally grow, into natural shrublands and grasslands. These plants are nature's carbon dioxide scrubbers.

It is a foggy autumn morning, and the upper canopy of my fir, hemlock, and cedar forest is barely visible. I take pride in knowing that the shrouded, thick foliage is helping fight climate change. It is a remarkable battle, technically called "photosynthesis," that my trees are waging against carbon dioxide. During the day, using the power from sunlight, chloroplasts deep in their needles drink in carbon dioxide and water and convert them to glucose and oxygen. Glucose is a sugar that plants use to fuel their life processes; extra glucose that trees don't use is converted into wood. The oxygen is released back into the atmosphere to be breathed by us and other animals.

It takes a lot of trees to have a meaningful effect on atmospheric carbon dioxide. In the United States, for example, our vast national forests store about 16 percent of the carbon we annually produce. All the street trees in U.S. cities together store only less than 1 percent of our carbon emissions. Still, the 360 metric tons of carbon that street trees store won't warm our climate. What if we converted half of subirdia's lawns to trees? How much carbon could they pull out of the air? I figured my backyard was a good place to start answering this question, and I knew just the person to help me determine the carbon capacity of my trees.

Dr. Dave Peterson is an ecologist with a passion for forests and their nemeses: fire, bugs, and climate change. As a member of the Intergovernmental

Panel on Climate Change, he contributed to the reports of the panel, which was awarded the Nobel Peace Prize in 2007. Dave and I taught together for years at the University of Washington, so naturally I turned to him for advice on how I might calculate the effect of subirdia's trees on carbon dioxide levels. Dave sent me a mound of equations that I could use to calculate the amount of living matter held within the trees of my yard and from that the flux of carbon in my part of subirdia.

It turns out that it is simple to determine the rough amount of carbon held in a tree. About half of the total tree mass is carbon, and mass can be calculated for a particular tree species simply by measuring the tree's diameter. Nearly all of my trees are Douglas-fir, so Dave sent me the equation that converts a fir's diameter into its aboveground biomass, the weight of its stump, trunk, and branches. Cool! I hit the yard with a measuring tape in hand and have weighed more than three thousand pounds and held about fifteen hundred pounds of carbon. Amassing this carbon doesn't happen quickly; each of my trees is about seventy years old. But the process is steady.

I got a feel for my trees' annual growth and carbon storage by measuring their change in size over the past two decades. Each tree, of course, carries this information within it, in its annual growth rings. So, all I had to do was take a small core of wood from my trees, a standard forestry technique, and examine the rings. As I counted and measured them, I discovered that for each of the past twenty years my trees increased their waistlines by a bit more than a tenth of an inch. As they fattened up on solar energy, they gained just over fifty pounds in dry mass per year and sucked nearly thirty pounds of carbon out of the atmosphere. To get that amount of carbon, each tree would have photosynthesized roughly one hundred pounds of carbon dioxide.

I may be off a tree or two, but I counted 362 of them on my two acres. Each of them is busy converting its annual one hundred pounds of carbon dioxide into a tenth of an inch of wood. And all that photosynthesis adds up. In total, my trees hold about 570,000 pounds of carbon and annually store an additional 10,000 pounds. And that's just the aboveground carbon; 20 percent to 30 percent more is stored belowground, in the soil and roots of my forest.

All that carbon storage sounds impressive, but how does it compare to the carbon my wife and I emit as we power our house, drive our cars, and live average American lives? I expected our carbon emissions to dwarf our trees' carbon consumption. But I was wrong! Using an online carbon calculator, I learned that our lifestyle produces just under thirty thousand pounds of carbon dioxide per year—equivalent to seventy-nine hundred pounds of carbon. Most of this comes from the natural gas we use for heat, though a substantial part also comes from our use of household electricity, gasoline for our cars, and airplane travel. The good news is that it is less than the amount of carbon that our trees annually sequester. Our yard is a carbon sink! In truth, our carbon emissions are greater still; our two daughters also travel extensively, but at least on a local level our trees are doing their part to alleviate our conversion of fossil fuels to climate-changing pollution.

Now, think about the effect that we could collectively have on carbon if only half of America's turflands—twenty million acres—could do what my trees do. Turf consumes carbon just as does any plant, but unlike trees, grass doesn't put on an annual layer of wood. Its use of carbon dioxide increases the soil's carbon stores. Estimates from 2003 suggest that an acre of U.S. grass annually adds about eighteen hundred pounds of carbon to the soil, and it continues to do so for about thirty years after it is planted. An acre of my trees did about three times that amount, in aboveground carbon storage, and they have been doing so for seventy years. Their annual storage capacity will slow as they age, probably in a century or so, but their overall capacity to buffer our climate is magnitudes beyond what grass could ever hope to do. Still, if we

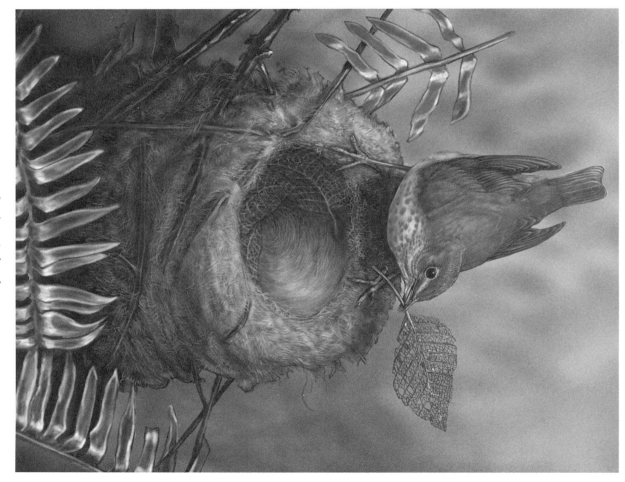

Swainson's thrush

use our national forests as a gauge, then an additional twenty million acres of trees could absorb another 1 percent to 2 percent of our national carbon emissions. In this accounting, every little bit helps!

Realizing what trees are doing for birds and climate helps people understand how their actions affect their ecosystem. Understanding can build respect, perhaps allowing some subirdia residents to overcome a fear of snags, frustration with blocked views, or the drudgery of cleaning needles and leaves from their gutters. Trees and the understory beyond grass that they support are the foundations of subirdia, and a sponge that helps us clean some of the ecological damage caused by an addiction to coal, oil, and natural gas.

As we develop more biologically diverse cities, we will be confronted with many difficult ecological lessons. Today, as a few species attain dense populations because of our inattention to the importance of diversity, many residents of subirdia are battling, rather than embracing, nature. The numbers of suburban deer, for example, have skyrocketed because of abundant food and few predators. Now they pose life-threatening hazards to drivers in the eastern half of North America, their hunger challenges gardeners and ground-nesting birds, and their ticks act as vectors for diseases that sicken some and scare many more.

To me, broadening our conservation ethic means that we must also learn to recognize and accept that in cities some wildlife that we cherish will also need to be humanely managed. Without their natural predators, abundant grazers, such as deer and geese, may need to be culled. Such decisions cannot be made in haste, nor should we avert our eyes from their consequences. Simply hiring a hunter and washing our hands of the problem teaches an incomplete lesson. Rather, I would encourage neighbors to work together to determine whether the health of their land would be improved by controlling overgrazing, and then put the result of the cull, done professionally, to good use. Suburban herds can offer residents and local charities high-quality, locally sourced food and leather. Su-birdia, we may learn, is able to promote health, aesthetic, economic, and utilitarian values. A diversity of returns may motivate a greater number of residents to conserve and foster a healthy ecosystem than any one factor alone is capable of doing.

The myriad services that our ecosystem provides are understood through observation, experience, and study. Citizen science programs offer an enjoyable way to learn, while collecting information essential to conservation. In Washington, D.C., for example, several hundred families work with biologists from the Smithsonian Institution to capture, band, and monitor birds living in their backyards. Citizens extend the scientists' eyes into the private habitats and habits of birds that live in urban, suburban, and rural settings. Scientists teach the citizens how to observe, record, and organize data detailing bird behavior, nest locations, and the fate of each breeding attempt. By working together, scientists and citizens learn more about birds from each other than either could learn by working alone. And the birds benefit as homeowners take responsible actions, such as planting shrubs, to benefit their feathered tenants. Beyond increasing their scientific literacy, resident citizen scientists increase their appreciation for wildlife generally, and, as they become more aware of life and its needs in their backyards, they increase their sense of place.

Citizen scientists, what some call amateurs, have long played an important role in discovery. On my bookshelf rests the work of several citizen scientists. Included are recent journals of observed crow behavior and older journals of exploration. My favorite and oft consulted was written by Meriwether Lewis and William Clark, citizen scientists dispatched in the early

1800s by President Thomas Jefferson, himself an accomplished citizen scientist, to learn about the western United States.

If you feel like carrying on the proud tradition pioneered by the likes of Lewis and Clark, you have many options. The Public Library of Science's blog CitizenSci includes a searchable database of more than five hundred projects just waiting for your assistance. Your help is needed to monitor everything from the squirrels, bats, and bees in your backyard to the microscopic life within your house. You could join a long-term project, helping assess the breeding success of powerful owls in Australia or the distribution and conservation needs of British birds, or spend as little as fifteen minutes helping count butterflies across the United Kingdom for the Big Count. If you have a day to spare during the holidays, you could join other amateur bird watchers on an Audubon Society Christmas Bird Count. These holiday counts, which were started more than a century ago, are held in most North American cities, and even some Central American ones. Being a citizen scientist is good for your brain and your ecosystem. The knowledge you gain informs science and policy, but more important, you will gain wisdom that can help guide your actions.

It is not easy to make the sacrifices that are required of us so that Earth is more able to support the other species that call the planet home. I fear that without a strong connection to nature, people will not even consider making sacrifices. Connecting with nature in the city may give future generations the strength to redefine progress to include measures of ecological and social quality in addition to the economic ones so relied upon today. A society better connected with its ecology might demand quarterly accountings of environmental indicators in addition to market indicators. They might push their political leaders to address the State of Nature along with the State of the City or the State of the Union. Tracking our natural world provides the feedback we need to gauge our effects and find the motivation to reduce them as needed for the sake of other species.

Watching the ecological struggles that occur in our yards, parks, and workplaces builds wonder that enables restraint. Few who watch young woodpeckers cry for food from the protection of their nest cavity would cut down a snag. Those who receive daily deliveries of dead warblers on their porch from a free-roaming cat would find it hard to support a local spay-neuter-return policy. Holding a tiny hummingbird, stunned after a recent collision with a window, encourages us to reconsider feeder placement and window transparency. Experience shapes our ethics and actions. If experience no longer includes nature, then our ethics cannot reflect the full needs of our natural world. Our interaction with nature is reciprocal—as we affect it, it affects us. Strengthening our place in the city's ecological web builds resilience to change and allows us to coexist with a wonderful diversity of life. Cutting our ties to the web is like cutting the belay line climbers rely upon as they stretch for a distant handhold. As we stretch to live within a rapidly changing world, are we ready to gamble on an unprotected, solo climb?

My hope is that we forego a lone ascent and instead seek creative ways to coexist with a wide diversity of birds and other creatures. Sacrificing to do this keeps nature intact and supports its influence, from our cities to their hinterlands. Facing our urban future in collaboration with our wild neighbors celebrates our appreciation for their intrinsic values as well as for the many ways that they improve our own beings. A lasting celebration increases the world's capacity to adjust to environmental change and quenches our thirst to remain part of nature.

Notes

Preface

In 2008 the United Nations reported that more than half of the world's population was urban. In 2011, 3.6 billion people lived in urban areas, a figure that was projected to increase to 6.3 billion in 2050. Virtually all of the world's population growth from 2011 is expected to be concentrated in urban areas, and migration of people from rural to urban homes is also expected to increase during this time. By 2050 it is projected that 6.3 of 9.3 billion people (68 percent) will live in urban areas (United Nations 2012).

Chiappe and Dyke (2001) describe the toothed birds of the Mesozoic; Longrich et al. (2011) link their extinction to asteroid impact.

Baillie and Groombridge (1996), updated regularly at http://www.iucnredlist.org, quantify bird extinctions and endangerment. Pimm et al. (1995), Czech and Krausman (1997), and Chapin et al. (2000) discuss factors putting birds at risk.

San Francisco's parrots are chronicled by Bittner (2004) and updated at Bittner's blog "Views from a Hill," http://markbittner.wordpress.com.

Peter Meffert has studied the wheatear in Berlin (Meffert et al. 2012).

The study of urban ecosystems has increased but remains a small percentage of scientific inquiry (Magle et al. 2012). Overviews of the field are presented in McDonnell and Pickett (1993), Marzluff et al. (2001), Berkowitz et al. (2003), Alberti (2008), Marzluff et al. (2008), McDonnell et al. (2009), Douglas et al. (2011), Endlicher et al. (2011), and Lepczyk and Warren (2012).

Aldo Leopold included his statement about land ethic in *A Sand County Almanac* (1949).

Turner et al. (2004) and Miller (2005) synthesize the ramifications of an urban population disconnecting from nature. Kellert (1999, 2012) synthesizes the ways nature informs human values in urban systems. Louv (2005) details the consequences of separating children from nature.

Chapter 1. Home Turf

Marsh (1864) wrote about humans in ecosystems, foreshadowing the field of urban ecology.

Richards (1990), Daniels (1999), Dwyer et al. (2000), Marzluff and Hamel (2001), Leu et al. (2008), Angel et al. (2011), Sterba (2012), and U.S. Environmental Protection Agency (2013) review the percentage of land on Earth that is occupied by urban and agricultural areas.

Wilson et al. (2012) define micropolitan and metropolitan.

Dwyer et al. (2000) estimated that there are seventy-five billion trees in U.S. metropolitan areas.

Richard Forman (2008) discussed the urban tsunami.

Early city populations were estimated by Berry (1990) and Gibson (2000) and are summarized by Wikipedia at http://en.wikipedia.org/wiki/List_of_largest_cities_through out_history. Estimates of populations in the Basin of Mexico are from Seis Ciudades Antiguas de Mesoamérica at http://www.cultura.inah.gob.mx/seisciudades/index.php ?option=com_content&view=article&id=5&Itemid=6.

United Nations (2012) defines urban population percentages and projections. United Nations (2007) projects an increase in moderately sized (less than five hundred thousand people) cities.

Daniels (1999) defines exurbia.

Modern city populations are tabulated by the United Nations (2012) and Cox (2012) and illustrated in a special report by *Time Magazine* (http://content.time.com/time/spe cials/packages/0,28757,2097720,00.html).

The Housing Assistance Council (2011) discusses the U.S. census definition of urban and its application to my home (Maltby, WA).

Jerry Coyne (2009) discusses human evolution.

Schimel et al. (2013) review changes in ecological processes that occur in the Anthropocene.

Chapter 2. Finding Subirdia

Ecological changes in the city are detailed in McDonnell and Pickett (1993), Pickett et al. (2001), Adams et al. (2005), Kaye et al. (2006), Shochat et al. (2006), Marzluff et al. (2008), McDonnell et al. (2009), Endlicher et al. (2011); and Douglas et al. (2011).

Seattle's city forest was assessed by Lisa Ciecko and Mark Mead and based on data from 223 one-tenth-acre plots distributed across Seattle. Other urban forests are equally diverse. For example, sampling 441 front yards in Auckland, New Zealand, yielded 4,704 trees and large shrubs from five hundred taxa; about 20 percent were indigenous (Meurk et al. 2009). The floras of Zurich, Switzerland, and Bochum, Germany, have both increased steadily over the past century to now include more than one thousand species of plants (Rebele 1994). Berlin, Germany, is typical of many European urban floras in having roughly half native and half nonnative species (593 and 839 species, respectively; Kowarik 1995). Kowarik and Körner (2005) synthesize these and other changes in urban forests.

The downtowns of cities in the Americas (Indianapolis, Indiana; Boise, Idaho; Ketchikan, Alaska; New York; Alajuela, Costa Rica; and Zihuatanejo, Mexico), New Zealand (Auckland), and Europe (St. Andrews, Scotland; Berlin, Germany; Oslo, Norway) are rife with birds.

McKinney and Lockwood (1999), McKinney (2006), and Clergeau et al. (2006) discuss biotic homogenization in urban areas. Catterall (2009) points out that homogenization is mostly confined to extremely disturbed parts of the city.

I visited Yellowstone National Park 25–28 March 2013 and Central Park 2–4 April 2013. Historical records of bird sightings in Central Park and Yellowstone are available at http://www.nycaudubon.org/pdf/birds-cpark-doc-oficial.pdf and http://www.nps .gov/yell/naturescience/birds.htm

Marzluff (2005) reports survey results from city to exurbs in Seattle. We count all birds detected within fifty meters of our observation point (Donnelly and Marzluff 2004a, 2004b, 2006).

The history of urban sprawl and statistics are from Ewing (1994), Daniels (1999), Theobald (2005), Leu et al. (2008), and Sterba (2012).

Sprawl as a factor in endangered species listing in the United States is from Czech and Krausman (1997) and Czech et al. (2000). Redford and Richter (1999), Marzluff and Hamel (2001), McKinney (2002), and Ricketts and Imhoff (2003) review the effect of sprawl on biodiversity.

Architect Albert Pope (1996) recognized the potential of suburbia and exurbia to contribute new cultural and natural identities to the city. Kunstler (1993) writes about "the geography of nowhere."

I use the 2006 National Land Cover Database to map subirdia (Fry et al. 2011) by highlighting areas classified as developed open space, low-intensity development, and high-intensity development (together this identifies areas with up to 80 percent impervious, or built, land cover).

Marzluff (2001), Chace and Walsh (2006), and Paker et al. (2013) evaluate the effects of urbanization on birds.

Gaston and Gaston (2011) consider biodiversity in English gardens.

Molino and Sabatier (2001) report on tropical tree diversity.

Blair (1996, 2001), Pennington and Blair (2011, 2012), Bock et al. (2008), Catterall (2009), Sewell and Catterwall (1998), Natuhara and Hashimoto (2009), MacGregor-Fors (2008), MacGregor-Fors et al. (2012), and Yan (2013) studied bird diversity along gradients of urbanization in California, Ohio, Arizona, Australia, Japan, Phoenix, Mexico, and Singapore, respectively. Taratalos et al. (2007) correlated British bird diversity with housing density.

Reviews of a variety of gradient studies that vary in findings and approaches can be found in Marzluff (2001) and Chace and Walsh (2006). Both papers refer to examples from desert cities where increased water directly, and indirectly through the diversity of vegetation it spawns, increases bird diversity.

Differences observed along various urban gradients likely result from differences in field techniques. An important study by Cam et al. (2000) shows how the measure of urbanization can affect the relationship one observes between urbanization intensity and biological diversity. When one measures the relative degree of heterogeneity that urbanization creates, one finds this measure to correlate positively with biological diversity and to peak at intermediate levels of urbanization. As more studies measure urbanization's effect on the diversity of remaining land cover, I expect more will show a peak in biological diversity at intermediate levels of urbanization, as we found in Seattle.

Jeff Norris's Ph.D. dissertation was published in 2011 by the University of Missouri–St. Louis.

Even if the species are not exactly the same, the types of species tend to become similar among various urban sites. This "functional homogenization" was described by Devictor et al. (2007) but is to be expected because urban areas provide similar resources and therefore favor similar types of birds around the world, notably species with tolerant, generalized resource needs.

Chapter 3. A Child's Question

I went birding in Lawrence, Kansas, on 12 April, 2012.

Rodewald (2012) documents the increase in Amur honeysuckle in suburban Ohio and its effects on the composition and productivity of birds. Avoiders generally decline with increasing honeysuckle, and when they nest within it, their productivity is low. Adapters increase with honeysuckle coverage, and their nesting success is less affected by its occurrence. Cardinals, in particular, increase with honeysuckle, and nesting within it reduces their breeding productivity only early in the season (Rodewald et al. 2010).

Neotropical migrant songbirds may be especially sensitive to development, because many require extensive canopy, shrub, or ground cover for nesting in North America (Friesen et al. 1995, Greenberg and Marra 2005). Neotropical birds that migrate to the United States and Canada to breed also avoid urban areas in Mexico where they overwinter (MacGregor-Fors et al. 2010). But in dry wintering areas, such as Phoenix, Arizona, the presence of riparian vegetation in well-watered suburbs provides important habitats for resident and migrant birds (Rosenberg et al. 1987).

In the late 1800s, cardinals were shy and confined to extensive woods (Burroughs 1871). In my visit to Kansas, as well as in a detailed study in Texas, cardinals are now abundant in suburban settings (Gehlbach 2005).

Aldrich and Coffin (1980) chronicle the change in avifauna around Lake Barcroft. As a fellow of the American Ornithologists' Union, Aldrich was honored with a memorial about his life (Banks 1997). Gehlbach (2005) engaged in a similar long-term study in Texas that lasted thirty-nine years.

Details on the technique of territory mapping known as "spot mapping" can be found in Kendeigh (1944) and Ralph et al. (1993).

Cramp (Cramp 1980, Cramp and Tomlins 1966) detailed the change in London's birds. The London bird atlas project (Woodward and Arnold 2012) brings Cramp's studies up to date.

Recher and Serventy (1991) documented changes in Kings Park, Perth, Australia.

Blair (1996) coined the terms "avoider," "exploiter," and "adapter."

Er et al. (2005) discuss bird extinctions in Vancouver, and Melles (2000) and Melles et al. (2003) surveyed Vancouver birds along a gradient of urbanization.

I used the U.S. Geological Survey Breeding Bird Survey (BBS) website to calculate changes in Seattle birds from 1966 to 2011 at larger scales (https://www.pwrc.usgs.gov /bbs/index.cfm?CFID=1154495&CFTOKEN=2a3c3ec0ee8a2b7a-D83991FA

-0821–9F12–86BC5DEAC3FAD622). The analysis section of the BBS website allows users to statistically analyze species trends for each state, region, or the nation as a whole. I assessed trends within the western region. Project FeederWatch also has an online database that allows you to estimate changes through time within various regions of North America. I used this to determine changes for the North Pacific Region, which includes British Columbia, Washington, and Oregon (http://www.birds.cornell.edu /pfw/DataRetrieval/rrendgraphs/index.html). The trends reported by Project Feeder-Watch are not based on the rigorous statistical analysis that the BBS provides, but they index relative changes in the percentage of feeders that were reported to have a particular bird visit it in a particular year (1985–2010) as well as the average number of birds seen at one time at feeders for a particular year.

House finch populations have declined in the eastern United States as a result of conjunctivitis (Hochachka and Dhondt 2000). You can track and report the spread of this disease and learn how to care for your feeders to reduce its spread at the Cornell Lab of Ornithology's House Finch Disease Survey, http://www.birds.cornell.edu/hofi /index.html.

House sparrow numbers have declined in Europe, Australia, North America, and elsewhere because of the decline of small-scale agriculture, removal of suitable urban habitat, and increased predation by raptors, such as the Eurasian sparrowhawk (Bell et al. 2010, Shaw et al. 2008)

It is a common feature of all ecosystems to have a few very common bird species and many very rare ones. In highly urbanized areas, such as the downtown core of a large city, this common pattern is accentuated as a few species dominate the ecological community and the overall number of species, and rare ones in particular, is diminished. Shochat et al. (2010) demonstrated this for birds and spiders in Phoenix, Arizona, and Baltimore, Maryland. Moreover, because successful urban birds can become extremely abundant, the larger trophic structure of cities may become overly influenced by avian predators, which may control numbers of herbivorous insects (Faeth et al. 2005). In the less densely settled suburbs of Seattle, trophic structure is typical for ecological systems, and the relative abundance of dominant species and their relationship to rare species (the slope of the species-abundance curve) is the same in developed and reserved portions of our study area.

The landscape surrounding our Seattle study areas has changed dramatically in the past twenty-five years as forests and farms have been converted to suburban neighborhoods (Robinson et al. 2005). Jack DeLap measured the composition of seven developments over the time period 1998–2010 that averaged 227 hectares (561 acres) in area.

Jack bounded his analysis by the area within 500 meters (about 1,600 feet) of each of eight bird survey points within each neighborhood.

Laura Farwell's study is detailed in Farwell and Marzluff (2012). In Australian urban landscapes the noisy miner is a bully similar to the Bewick's wren (Catterall 2009; Kath et al. 2009; see Chapter 4).

To analyze movements of banded birds in our study required that we spot map their territories in at least two years. We had sufficient data to do this only for adults of four species (song sparrow, spotted towhee, Swainson's thrush, and Pacific wren). I calculated the distances between territory site centers for each pair of years we mapped an individual's territory. This calculation resulted in 17, 50, 3, and 14 comparisons in reserves for sparrows, towhees, thrushes, and wrens, respectively; 54, 14, 1, and 2 comparisons for the same species in developments; and 103, 102, 10, and 5 comparisons in sites undergoing active development.

Keil (2005) notes the importance of urban woodlots to serve as adventure sites for children.

Louv (2005) provides an early synthesis of the effects of children's increasing separation from nature. Wells and Lekies (2012) synthesize the recent literature on how separation reduces childrens' concentration, cognitive function, physical condition, and health. Kellert and Wilson (1995) developed the biophilia hypothesis. Kellert (2012) reviews the connections and implications of people interacting with nature in the modern world.

Chapter 4. A Shared Web

Faeth et al. (2005) discuss the trophic structure of urban systems, especially noting the importance of human subsidies.

Food web complexity and redundancy are thought to increase sustainability (Polis and Strong 1996; Paine 1966, 1980; Loreau 2000; Naeem 2002). Facilitation is a stabilizing force for this reason (Hacker and Gaines 1997, Stachowicz 2001, Bruno et al. 2003).

Paine (1966) developed the concept of a keystone species. Predators are often keystone species, and their prey may increase greatly and affect ecosystem stability in their absence (Terborgh et al. 2001).

The three common species of *Myotis* bats in western Washington—California myotis, little brown myotis, and western long-eared myotis—are best distinguished by their barely audible vocalizations. Without a specialized "bat detector," I could not tell them apart.

Woodpeckers select nest trees on the basis of subtle cues to their hardness (Schepps et al. 1999), especially the presence of heartwood decay (Harestad and Keisker 1989). Secondary cavity nesters often require soft wood (Steeger and Hitchcock 1998), but some that nest in especially soft snags are frequently preyed upon (Christman and Dhondt 1997).

Blewett and Marzluff (2005) quantified snag density in urban forests.

Kathy Martin (Martin and Eadie 1999, Martin et al. 2004, Cockle et al. 2012) illustrated the details of a cavity nest web and demonstrated the relationship between woodpecker abundance and the richness of bird communities (Drever et al. 2008). Dobkin et al. (1995) detailed a particular nest cavity web in the Rocky Mountains.

Northern flickers eat primarily ants and beetles (Moore 1995).

Jorge Tomasevic's dissertation on urban pileated woodpeckers is in preparation.

Barbara Clucas's and Sonya Kübler's results are published in Clucas et al. (2011) and Clucas and Marzluff (2011, 2012).

Koenig (2003) reviewed the potential negative effects of starlings on cavity nesters and found little influence. Ingold (1996) and Blewett and Marzluff (2005) reported on usurpation of flicker nests by starlings. Ingold (1989, 1994) studied the negative effects of starlings on red-bellied woodpeckers and red-headed woodpeckers.

Hudson reported starling nesting on a building in 1893 (Clews 2006).

Raptor responses to urban environments, including increases in peregrines across the United States and their diverse prey, are detailed in Bird et al. (1996). Particular shifts in urban tawny owl and kestrel diets are from Goszczyński et al. (1993) and Kübler et al. (2005). Sol et al. (2013) describe many other changes in behavior as species invade the city.

Stan Rullman's Ph.D. dissertation (2012) includes his research on Cooper's hawks. Rullman and Marzluff (2014) report on the change in hawks and owls across the urban gradient, including Seattle.

Urban nesting Cooper's hawks have also been studied in Arizona (Boal and Mannan 1998) and Wisconsin (Rosenfeld et al. 1996).

In Norwegian forests another sort of complex relationship between predator and prey plays out (Slagsvold 1980). Fieldfares are thrushes very similar in size to the American robin; both belong to the same genus, *Turdus*. Fieldfares live up to this scientific name. They nest in colonies and cooperate to protect their nests by ganging up on predators such as hooded crows and Tengmalm's owls and bombing them with feces. Would-be predators stay clear, and as a result other small birds crowd into fieldfare colonies and enjoy high nesting success. High densities of some birds such as the bramling, redwing, goldcrest, and willow tit in fieldfare colonies, however, appeared to exclude others such as the coal tit, bullfinch, and European robin. As a result, although abundance was greater, diversity of songbirds was actually lower in fieldfare colonies than outside of them. It seems that crows in Norway diversify small bird communities by reducing the density of especially common and competitive birds. The extent to which this occurs in suburban settings has not been investigated.

Grant and Grant (2011) describe character displacement in Galapagos finches.

Connell (1980) introduces the ghost of competition past, and Wiens (1977) discusses the role of ecological crunches in competitive interactions.

Kath et al. (2009) studied noisy miners in Crows Nest Shire.

Sedláček et al. (2004) studied two European redstarts in the Czech Republic.

Black drongos were introduced to the Northern Mariana Islands, where they are suspected of harassing a number of small passerines and the endangered Mariana crow (U.S. Fish and Wildlife Service 2005).

Livezey (2009), Dugger et al. (2011), and Diller (2014) describe the competitive relationship between invading barred owls and resident spotted owls in the Pacific Northwest.

Dr. Al Sanford, a long-time resident of Socorro, New Mexico, provided notes on the changing dove community witnessed in this small town. Bled et al. (2011) provided the invasion history of Eurasian collared doves. The spread of Eurasian collared doves and their effects on other species of doves in Florida was documented using Project FeederWatch data by Bonter et al. (2010). Robertson and Schnapf (1987) describe the pyramiding behavior of Inca doves. Shochat et al. (2010) discuss how domination of urban ecosystems by successful species, such as doves and pigeons, may reduce resource availability to other species and therefore limit biodiversity.

Chapter 5. *The Fragile Nature of Subirdia*

Female bushtits are born with dark irises, which gradually lighten until fully yellow by age one month (Ervin 1975). The functional significance, other than to indicate sex in this species, is not known. Sloane (1996) and Bruce et al. (1996) report that bushtits breed monogamously and that pairs retain ties to their larger social flock throughout the breeding season.

In general, researchers report increased nest predation in urban areas (Chamberlain et al. 2009, Stracey and Robinson 2012) that particularly affects nests in proximity to yard interfaces (Marzluff and Restani 1999) and ground nesters (Kaisanlahti-Jokimäki et al. 2012), though not in all situations (Gering and Blair 1999, Marzluff et al. 2007, Ryder et al. 2010, Smith et al. 2012)

From 1998 to 2010, we found 1,661 nests; 47 percent (779) belonged to either robins (589) or Swainson's thrushes (190). The methods we use to survey diversity and demography of birds along the gradient of urbanization in Seattle are detailed in Donnelly and Marzluff (2004a, 2004b, 2006).

American robins laid on average 2.89 eggs per clutch and fledged 2.47–2.64 young, depending on the landscape. Swainson's thrush clutches averaged 3.33 eggs, and they fledged 2.75–3.06 young per nesting attempt. Swainson's thrushes fledged more young per attempt in reserves than did robins (3.06 versus 2.63).

Twenty-one percent of robins fledge two broods in changing and reserve landscapes; 15 percent of robins do likewise in neighborhoods. In contrast, 0.4 percent of Swainson's thrushes fledge two broods, and only in reserves. As a result, the season-long production of fledglings by both thrushes was equal across landscape types. Fledgling female birds (assuming a fifty-fifty sex ratio) per breeding female in changing, developed, and reserve study sites, respectively, are as follows (mean to upper 95 percent confidence interval): American robin, 0.87–1.05, 0.61–0.85, 0.87–1.08; Swainson's thrush, 0.60–0.86, 0.63–1.35, 0.91–1.19.

The number of young fledged per nesting attempt by our other focal species is as follows: Bewick's wren, 3.47–3.63; dark-eyed junco, 2.81–3.36; song sparrow, 2.88–3.12; spotted towhee, 2.2–2.54; Pacific wren, 3.11–3.49. Song sparrows and juncos fledged two broods per year from 6 percent to 15 percent of the time, depending on landscape; double production was rare in the other species. Total annual productivity (female fledglings per female breeder, as above for thrushes) in changing, developed, and reserve study sites, respectively, averaged as follows: Bewick's wren, 1.09–1.64, 1.15–1.73, 1.30–2.03); dark-eyed junco, 1.33–1.73, 1.08–1.56, 0.97–1.35; song sparrow, 1.23–1.56, 1.17–1.48, 1.07–1.58; spotted towhee, 0.85–1.09, 0.72–0.94, and 0.69–1.02; Pacific wren, 0.71–1.05, 0.82–1.21, 0.82–1.13.

Smith et al. (2012) have studied spotted towhees in Portland, Oregon.

We found a total of 344 nests of primary and secondary cavity nesters. Most woodpeckers (72 percent of 48 nests) and secondary cavity nesters (89 percent of 266 nests) fledged young. Secondary cavity nesters, both native and nonnative species, were more successful in human-derived structures (nest boxes, buildings, light poles, etc.) than in natural cavities (mostly in snags). Nearly every (156 of 159, or 94.9 percent) house sparrow and starling nest we watched succeeded.

Chamberlain et al. (2009) synthesize productivity studies across various urban settings.

Fuller et al. (2012) tabulated and summarized studies that estimated participation, investment, and types of seed used in bird feeding by Europeans, North Americans, and Australians. These authors also related bird feeding to human density and other aspects of the built environment, assessed the provision of nest boxes as well as feeders, and related both types of supplements to bird abundance, distribution, and diversity. Dunn and Tessaglia-Hymes (1999) provided more extensive analysis of bird feeding in the United States, including analysis of bird use of feeders as reported by Cornell University's Program FeederWatch. Lepczyk et al. (2012) compared the provision of resources (food and water) among regions of the United States. Clucas and Marzluff (2011, 2012) and Clucas et al. (2011) surveyed participation in bird feeding and nest box provisioning by citizens in Berlin, Germany, and Seattle.

Sterba (2012) provides an interesting history of bird feeding in the United States.

Densities of individual species and diversity of bird species are both positively correlated with feeders and the diversity of foods provided (Horn 1999, Jokimäki et al. 1996, Clucas and Marzluff 2011 and 2012, McCaffrey et al. 2012, Rotenberg et al. 2012, Vargo et al. 2012).

Estimates of U.S. grain shipments to Africa is from Grainnet, 17 July 2000: http:// www.grainnet.com/articles/USDA_Ag_Secretary_Glickman_Announces_Additional _Food_Donations_For_Africa__Plans_to_Visit_Three_African_Countries-7541 .html.

Dunn and Tessaglia-Hymes (1999) report that the distributions of some species, such as northern cardinals, tufted titmice, and Anna's hummingbirds, have expanded northward in response to bird feeders.

Schoech and Bowman (2001, 2003) have studied the use of feeders by suburban Florida scrub-jays.

Black-capped chickadee response to bird feeders has been especially well studied, including effects of feeders on survival, breeding density, breeding condition, and use of other foods (Brittingham and Temple 1988, 1992a, 1992b, Grubb and Cimprich 1990).

Robb et al. (2008b) review many studies that assess the effect of feeders on laying date, clutch size, egg size, incubation time, hatching success, chick growth, fledging success, overwinter survival and condition, behavior, and changes in the distribution of populations.

Increased reproductive output of pinyon jays and blue tits fed over winter was reported by Marzluff and Balda (1992) and Robb et al. (2008b), respectively.

Birds with access to feeders often breed about one week earlier than those without such access (Robb et al. 2008a), which may cause chicks to be fledged at times when natural foods they require are rare (Schoech and Bowman 2001).

The British Trust for Ornithology recommends feeding birds year-round (Toms 2003).

Loss et al. (2013) estimated that there are 10–20 billion birds in the contiguous United States and that the median number killed by cats was 2.4 billion. The detailed appendices in this report also provide estimates of American cats and cat owners. Calvert et al. (2013) undertook a similar study in Canada. Jarvis (2011a, 2011b) provide estimates of cat abundance and effect in Europe and Australia. Baker et al. (2005) measured cat consumption of sparrows, dunnocks, and robins in Bristol. Bonnington et al. (2013) estimated the influence of cats on parental nest attendance and subsequent nest success.

The actions of cat owners to reduce predation are largely ineffective (Williams 2009). Keeping cats as pets inside greatly reduces common causes of death such as being struck by a car, poisoned, or contracting a fatal disease. Reducing such accidental deaths results in a lifespan of twenty years on average for indoor cats versus ten years on average for outdoor cats (Lacheretz et al. 2002).

Catching and banding birds requires several permits and special training. My research requires a federal banding permit, a state collecting permit, and approval from the University of Washington's Animal Care and Use Committee.

Few birds can be reliably aged without banding. The birds I studied could be classified as hatch year (born the summer that I caught them), after hatch year (one year old when captured), or after second year (at least two years old when captured) by a combination of plumage characteristics (Pyle 1997).

Dave Oleyar's thesis was published in 2011. Dave calculated annual survivorship on the basis of our mark and recapture efforts using Program Mark (White and Burnham 1999) and a model that included age and the interaction between species and landscape. Annual survivorship rates from 2002 to 2009 for each species in changing, developed, and reserve landscapes, respectively, are as follows (mean plus upper 95 percent confidence interval). Adults: American robin, 0.63–0.81, 0.44–0.88, 0.71–0.91; Bewick's wren, 0.53–0.70, 0.52–0.76, 0.58–0.78; dark-eyed junco, 0.45–0.58, 0.62–0.83, 0.54–0.75; song sparrow, 0.57–0.66, 0.63–0.72, 0.54–0.64; spotted towhee, 0.55–0.64, 0.46–0.63, 0.59–0.70; Swainson's thrush, 0.58–0.69, 0.36–0.63, 0.56–0.70; and Pacific wren, 0.52–0.69, 0–1, 0.48–0.66. Juveniles: American robin, 0.26–0.48, 0.14–0.63, 0.34–0.69; Bewick's wren, 0.19–0.33, 0.18–0.41, 0.23–0.43; dark-eyed junco, 0.14–0.23, 0.17–0.32, 0.20–0.39; song sparrow, 0.22–0.31, 0.26–0.37, 0.19–0.29; spotted towhee, 0.20–0.29, 0.15–0.27, 0.23–0.34; Swainson's thrush, 0.22–0.34, 0.11–0.27, 0.21–0.34; and Pacific wren, 0.19–0.32, 0.17–0.31, 0.16–0.30.

Kara Whittaker's thesis was published in 2007, with two individual chapters published as Whittaker and Marzluff (2009, 2012). Her efforts indicated that the survival of juvenile songbirds in changing, developed, and reserve landscapes, respectively, was as follows: song sparrow, 0.48, 0.46, 0.59; spotted towhee, 0.33, 0.31, 0.35; and Swainson's thrush, 0.48, 0.48, 0.48.

Mary McFadzen studied postfledging survival in prairie falcons and documented the effect of great horned owls on them (McFadzen and Marzluff 1996).

Dave Oleyar's estimates of landscape-specific annual rates of population change (lambda) are as follows for changing, developed, and reserve study sites, respectively (means plus upper 95 percent confidence interval: American robin, 0.85–1.3, 0.53–1.42, 1.01–1.65; Bewick's wren, 0.73–1.24, 0.73–1.46, 0.88–1.65; dark-eyed junco, 0.64–0.99, 0.81–1.32, 0.73–1.28; song sparrow, 0.84–1.41, 0.94–1.1, 0.74–1.43; spotted towhee, 0.72–1, 0.57–0.92, 0.75–1.05; Swainson's thrush, 0.71–1.09, 0.43–1.28, 0.75–1.27; and Pacific wren, 0.65–1.02, 0–1, 0.61–0.99.

Amanda Rodewald (Rodewald and Shustack 2008) studied Acadian flycatchers and northern cardinals in Ohio.

Thomas Unfried's dissertation was published in 2009.

Stochastic, or random, variation leads to extinction in small populations (Lande et al. 2003). Dave Oleyar (2012) used the program RAMAS (Akçakaya 2002) to estimate the likelihood of extinction for each songbird we studied.

Stacey and Taper (1992) studied the dynamics of acorn woodpeckers.

You can report banded birds you encounter to the North American Bird Banding Program, Bird Banding Laboratory, at http://www.pwrc.usgs.gov/bbl/.

Chapter 6. Where We Work and Play

The epigraph is from Cristol and Rodewald (2005). This material is reproduced with permission of John Wiley & Sons, Inc.

I observed blackbirds at the Woodinville, Washington, Costco store on 3, 28, and 31 January and 3 February 2013.

Horn (1968) described the colonial breeding system of Brewer's blackbirds.

Robbert Snep's dissertation, including the work at the Port of Antwerp, was published in 2009.

Land coverage, rate of growth, and basic statistics about golf courses are reported by Cristol and Rodewald (2005), Tanner and Gange (2005), Hodgkison et al. (2007a), Colding and Folke (2009), and Hudson and Bird (2009). Challenges of golf courses to birds are also discussed by all these authors. Rainwater et al. (1995) studied the specific effects of pesticides in South Carolina.

The role of golf courses to aid wetland loss was discussed by Merola-Zwartjes and DeLong (2005) and White and Main (2005).

I birded Balmoral Golf Course on 20 May 2012 and the ancient links at St. Andrews, Scotland, on 20 February 2013.

Declines in red-headed woodpeckers have been discussed by Ingold (1989) and Rodewald et al. (2005).

Diversity of birds on golf courses was tabulated in Queensland, Australia (Hodgkison et al. 2007a, Davis et al. 2012); Quebec, Canada (Hudson and Bird 2009); South Africa (Fox and Hockey 2007); Italy (Sorace and Visentin 2007); and the United States (White and Main 2005, Jones et al. 2005, Merola-Zwartjes and DeLong 2005, LeClerc and Cristol 2005, Porter et al. 2005, Smith et al. 2005, and Terman 1996).

Fox and Hockey (2007) documented bird loss with golf resort development in the Strandveld.

White et al. (2012) studied the response of white-breasted thrashers to golf resort development in St. Lucia.

Sauer and Droege (1990) evaluated general declines in eastern bluebirds. The response of eastern bluebirds to golf courses was studied by Stanback and Seifert (2005) in North Carolina and LeClerc et al. (2005), Knight and Swaddle (2007), Cornell et al. (2011), and Jackson et al. (2011, 2013) in Virginia.

Little and Sutton (2013) studied problems with geese on golf courses.

Terman (1996) studied the birds at Prairie Dunes. I spoke with current course manager Scott Nelson on 2 August 2013.

Donnelly and Marzluff (2004a) studied the relationship of Seattle forest birds to urban forest patch size.

Terman (1996), Cristol and Rodewald (2005), Porter et al. (2005), Colding (2007), Hodgkison et al. (2007a, 2007b), Colding and Folke (2009), Hudson and Bird (2009), and Hedblom and Söderström (2012) have discussed planning golf courses and other urban green spaces relative to surrounding landscape.

The evolutionary responses of cliff swallows to life above roadways was presented by Brown and Brown (2013).

Chapter 7. The Junco's Tail

The epigraph is from *The Evolution Explosion* by Stephen R. Palumbi. Copyright © 2001 by Stephen R. Palumbi. Used by permission of W. W. Norton & Company, Inc.

Darwin (1859) devoted the first chapter of *On the Origin of Species* to variety and change in domestic species, with a special section about domestic pigeons.

Thompson (1998), Hendry and Kinnison (1999, 2001), Kinnison and Hendry (2001), Palumbi (2001), and Stockwell et al. (2003) have evaluated contemporary, rapid evolution. I reviewed its potential to influence birds in urban environments (Marzluff 2012).

I assessed the juncos at the Burke Museum on 25 June 2013. The tag on the old specimen did not cite the collector, but Lucien Turner was collecting in Labrador during the year this bird was obtained. Most of his specimens were preserved in alcohol, which the Burke bird was not, so I am not positive he procured this bird. I measured the tails of adult male juncos and scored the number of white feathers on their right side following Yeh (2004). My results (not published) indicate significant differences between wildland samples ($n = 20$ birds from the Cascade Mountains and Canada during years 1891–1990, mean of 2.54 white tail feathers, SE = 0.07) and urban samples ($n = 24$ birds from western Washington cities during years 1910–1998, mean 2.12, SE = 0.05); a slight difference existed between thirteen urban juncos collected from 1910 to 1940 (mean 2.22, SE = 0.09) and nineteen collected (or measured during my field study) from 1988 to 2010 (mean 2.10, SE = 0.05).

Yeh (2004), Yeh and Price (2004), and Atwell et al. (2012) investigated rapid evolution of junco traits when juncos colonized San Diego. Walasz (1990) and Partecke et al. (2004, 2006a, 2006b) studied changes in European blackbirds.

Reznick and Ghalambor (2001), Price et al. (2003), and Yeh and Price (2004) discuss how phenotypic plasticity during colonization provides variability for natural selection to work on and may therefore promote speciation.

Gehlbach (1988) discussed adaptation to urban ecosystems by Eastern screech-owls, including differences in their color morphs. Baker and Moeed (1979) documented changes in mynas after they were introduced and spread throughout New Zealand.

House sparrows were the focus of Dick Johnston's research (Johnston and Selander 1964, 1971; Selander and Johnston 1967; Johnston et al. 1972). Blem (1974) studied the heat loss by sparrows of various proportions. The relationship of body size to climate is less obvious among sparrows of the far north and among those in New Zealand (Baker 1980). In the extreme north of North America (above 45°N), the size of sparrows declines with further increases in latitude, possibly reflecting more humid conditions or reduced food availability. In New Zealand, the largest sparrows are from the warm, humid north, suggesting that climate may sculpt sparrows differently, more slowly, or simply less than it appears to do in North America. In New Zealand, the vagaries of a small founding population may constrain adaptive changes in body size expressed in North America.

Shaw et al. (2008) and Bell et al. (2010) document declines in house sparrows.

Bishop and Cook (1980) provided evidence for industrial melanism as an evolutionary force in urban environments. Cook et al. (1986) documented increases in light morphs after coal use was reduced.

Murton et al. (1973, 1974) and Hetmański and Jarosiewicz (2008) speculate on the reproductive advantage of dark pigeons.

Palumbi (2001) details the evolution of resistance to antibiotics and pesticides. Haugen and Vøllestad (2001) document grayling evolution.

In Hawaii, 90–110 of 125–145 bird species have become extinct (Pimm et al. 1994). Marzluff (2012) and Marzluff and Angell (2005b) detail cultural evolution in urban settings.

Patricelli and Blickley (2006), Warren et al. (2006), and Slabbekoorn et al. (2007) summarize the changes in urban birdsong. Arroyo-Solís et al. (2013) experimentally confirm the earlier dawn chorus in urban birds. Lowry et al. (2012) confirm that noisy miners are noisier in urban settings. Luther and Derryberry (2012) document cultural evolution in white-crowned sparrow songs. Potvin et al. (2013) have studied Australian silvereyes. Marzluff and Angell (2005a, 2005b) proposed cultural coevolution.

Clucas et al. (2013) reported on the response of crows to gaze.

My ancestors are from the northern border of Italy. Johnston (1969), Fulgione et al. (2000), Dinetti (2007), and Hermansen et al. (2011) studied the role of hybridization in the evolution of Italian sparrows.

Terrill and Berthold (1990), Berthold et al. (1992), Bearhop et al. (2005), and Rolshausen et al. (2009) detail the evolution of a new migratory route in blackcaps. Marzluff and Angell (2005a) and Able and Belthoff (1998) discuss shifts in crow and house finch migrations.

Proppe et al. (2013) suggest that urban noise is important to the local extinction of some Edmonton birds and that it may increase the similarity in bird communities across many cities.

Lande (1976) and Gomulkiewicz and Holt (1995) discuss the importance of demographic stochasticity and random genetic drift to extinction. Marzluff (2012) considers the evolutionary process and species likely to adapt versus to become extinct in urban environments.

Local adaptation, which can be eroded by gene flow, may be important in urban evolution (Marzluff 2012), as it is in maintaining the adaptive potential of sea turtles (Stiebens et al. 2013).

Thomas Unfried's song sparrow investigation (Unfried et al. 2013) considered gene flow in an urban setting.

Trut (1999) considers the long-term fox-breeding experiments by Dmitry Belyaev and his colleagues.

Chapter 8. Beyond Birds

I surveyed the mammals of Yellowstone and Central Park 25–28 March and 2–4 April 2013, respectively.

Coyote sightings in Central Park and public reaction to them are reported by Stroud (2012). More bizarre sightings are reported on the NYC Parks website (http://www.nyc govparks.org).

Our small mammal surveys in Seattle occurred from 1998 to 2001. Brief results are published in Hansen et al. (2005). Rieu et al. (2012) and Cavia et al. (2009) studied the small mammals of Ohio and Buenos Aires, respectively.

Adams (1994) discussed common urban mammals, and other ecological aspects of cities, noting the increased density of small mammals such as squirrels in Lafayette Square. Hoffman and Gottschang (1977) studied suburban raccoons in Cincinnati.

Diets of coyotes were studied in California by Fedriani et al. (2001) and in Seattle by Quinn (1997). Fedriani et al. (2001) projected increases in suburban carnivores that utilize human foods. The consumption of cats by urban coyotes is a significant benefit for birds (Soulé et al. 1988).

Melbourne's flying foxes were studied by Parris and Hazell (2005). Their current status can be tracked at the Royal Botanic Gardens website, https://www.rbgsyd.nsw .gov.au/welcome/royal_botanic_garden/gardens_and_domain/wildlife/flying-foxes.

Kertson et al. (2011, 2013) study cougars in western Washington. Burdett et al. (2010) study cougars in southern California.

Sauvajot et al. (1998) studied small mammal and bird responses to suburban development of chaparral on the outskirts of Los Angeles.

Riley et al. (2003, 2006) tracked the movements of coyotes and bobcats with respect to the Ventura Highway in southern California to document the genetic effects of changed ranging patterns. Wandeler et al. (2003) noted that the genetic differences in urban Swiss foxes that were apparent immediately after they colonized Zurich are now eroding, because foxes are increasing and spreading throughout the city and surrounding country, thereby improving gene flow between formerly isolated populations.

Mitchell et al. (2008) detail the responses of reptiles and amphibians to urban development. Mitchell and Brown (2008) synthesize impacts and conservation measures needed to keep herps in urban settings. I updated the information on the loss of amphibians and reptiles from New York City, San Francisco, and Seattle by consulting state herpetological atlases (New York: http://www.dec.ny.gov/animals/277.html; California: http://www.californiaherps.com; Washington: www.dnr.wa.gov/nhp/refdesk /herp/speciesmain.html).

Sutherland and Stranko (2008) studied amphibians in Maryland.

I consulted with Dr. Stephen D. West and Amy Yanke to update and expand the information on Seattle's amphibians presented in Ostergaard et al. (2008).

The importance of wetland-terrestrial connectivity has been emphasized by the studies published in 2008 by Trenham and Cook, Windmiller et al., Egan and Paton, Clark et al., and Paloski. Paton et al. (2008) radio-tracked salamanders crossing suburban golf courses.

Andrews et al. (2008) and Schmidt and Zumbach (2008) reviewed the detriments of roads to herps. Hagood and Bartles (2008) and Kühnel (2008) document the use of road crossings by turtles and lizards.

Ferebe and Henry (2008) and Brisbin et al. (2008) studied Eastern box turtles.

Golf courses in Detroit, Michigan; Queensland, Australia; Tucson, Arizona; and New England were studied by Mifsud and Mifsud (2008), Hodgkison et al. (2007a), Smith et al. (2008), and Paton et al. (2008), respectively.

Scott et al. (2008) studied amphibians in seasonal wetlands on golf courses in South Carolina and Georgia. Giles et al. (2008) documented the use of urban Australian lakes by turtles.

Germaine and Wakeling (2001) studied lizard diversity along a gradient of urbanization that included Tucson, Arizona. Reptile diversity is also high on Tucson golf courses, but it declines with the age of the course (Smith et al. 2008). Conversely, older courses with established trees may in fact be more attractive to birds than newly created ones. Chapman and Underwood (2009) document the increase in urbanization along coasts, the relative lack of study in urban marine systems, and the important differences between urban effects on land and sea. Chapman et al. (2009) review the influence of cities on nearby marine environments. The effects of invasive species in marine and freshwater ecosystems (as well as terrestrial systems) are included in the book edited by Mooney and Hobbs (2000).

Francis and Schindler (2006, 2009) and Schindler et al. (2000) assess changes in North American lakeshores in response to urbanization and their effects on fish.

The effects of night lighting on insects were reviewed by Eisenbeis and Hänel (2009). Meyers et al. (2013) conducted experiments on light's effect on aquatic invertebrates.

McIntyre and Rango (2009) studied the insects of Phoenix. Johnson et al. (2012) focused on the black widows of Phoenix. Sattler et al. (2010) studied Swiss bee, spider, and bird diversity.

McIntyre and Hostetler (2001), Matteson and Langellotto (2010), and Yan (2013) studied urban bees and other pollinators.

McDonnell et al. (1997), Pickett et al. (2009), and Loss and Blair (2011) studied urban soils, worms, and their effects on nutrient cycling and birds. General information on invasive earthworms can be found at http://www.allaboutwildlife.com.

Whipple and Hoback (2012) surveyed carrion and dung beetles in New York City. I discussed the situation with Dr. Hoback on 19 August 2013, and he indicated that it is not known whether dung beetles bury goose feces, but they do prefer herbivore feces to the waste of other species.

Blair (2001), Natuhara and Hashimoto (2009), Hochuli et al. (2009), Matteson and Langellotto (2010), and Yan (2013) studied the responses of butterflies to urbanization.

Tarkhnishvili and Gokhelasuili (1996), Mitchell and Brown (2008), and Ishchenko and Mitchell (2008) discuss the moor frogs of Russia.

Lampe et al. (2012), Parris et al. (2009), and Slabbekoorn and Ripmeester (2008) studied the effects of road noise on grasshoppers, froglets, and great tits, respectively.

I synthesized the effect of roadkill on reptiles and amphibians from Mitchell and Brown (2008) and Andrews et al. (2008) and obtained roadkill statistics from Seiler (2003), U.S. Department of Transportation (2008), Sterba (2012), and Calvert et al. (2013).

Brown and Brown (2013) studied cliff swallow evolution.

Eddie the eagle's story can be read at http://seattletimes.com/html/localnews/2015800 831_eagle2m.html, "520 bridge's beloved bald eagle is struck and killed," by Jessie Van Berkel. An example of a blog that regularly features musings about Eddie is Union Bay Watch, http://unionbaywatch.blogspot.com/2012/06/life-after-eddie.html.

Mitchell and Brown (2008) synthesize the effects of persecution on snakes and frogs.

Sterba (2012) focuses on the problems animals cause in urban areas. Blaustein and Johnson (2003) summarize challenges to amphibians that lead to deformities and population declines. The influence of night lighting on reptiles and amphibians in urban settings is reviewed by Perry et al. (2008).

Barratt (1998) and Loss et al. (2013) assessed the effects of cats on small mammals and herps.

You can learn about the citizen efforts on Beekman Road at Friends of the East Brunswick Environmental Commission, http://www.friendsebec.com/salamandercrossing.html.

Frog logs and toad tunnels are described by Hagood and Bartles (2008), Mason (2008), and Schmidt and Zumbach (2008).

Chapter 9. Good Neighbors

The epigraph is from *A Sand County Almanac*, by Aldo Leopold (1977), © 1949, 1977 by Oxford University Press, Inc. By permission of Oxford University Press, USA.

Jack Ewing manages Hacienda Barú (http://www.haciendabaru.com/), and Marino Chacón Zuniga stewards the land owned by the Savegre Hotel (http://www.savegre.com/).

Bormann et al. (2001) provide a detailed history of the American lawn, including Olmstead's view of it, and the ecological implications of traditional lawns and Freedom Lawns. Robbins (2007) explores the influence of lawns on people.

Milesi et al. (2005) mapped the extent of turf in the United States.

The economic and ecological cost of lawns in the United States is discussed by Bill Chameides, "Stat Grok: Lawns by the Numbers," at http://www.huffingtonpost.com/bill-chameides/stat-grok-lawns-by-the-nu_b_115079.html.

Pollan's view of the shaggy lawn is quoted in the Yale group's book on the American lawn (Bormann et al. 2001).

The requirements of establishing a backyard wildlife sanctuary in Washington State can be found at Washington Department of Fish and Wildlife, "Living with Wildlife," http://wdfw.wa.gov/living/backyard. The National Wildlife Federations program is described at "Garden for Wildlife," http://www.nwf.org/How-to-Help/Garden-for-Wildlife/Certify-Your-Wildlife-Garden.aspx. The Humane Society of the United States' certification program is described at "Humane Backyard," http://www.humanesociety.org/animals/wild_neighbors/humane-backyard/humane-backyard.html.

Learn more about Randall Arendt at http://www.greenerprospects.com/index.html. Milder (2007) and Hostetler and Drake (2009) review the principles of conservation design and their implications for wildlife.

In the conservation design neighborhood I counted twenty-five species in 1998 and thirty-three in 2010. In contrast, the nearby forest reserve held twenty-three in 1998 and twenty-four in 2010, while the adjacent traditional neighborhood had twenty-nine in 1998 and thirty in 2010. In all years, the neighborhoods held substantially more individuals that did the reserve, but in both neighborhood styles I counted similar numbers of individual birds.

Dave Oleyar teamed up with students of economics and human well-being to determine that homeowner satisfaction and home value in our study area increased in conservation neighborhoods and other places in proximity to forests (Oleyar et al. 2008).

Increasing plant diversity, especially native shrub cover, is a key to improving the habitat quality of suburbs for birds (Knight 1990, Catterall et al. 1998, Donnelly and Marzluff 2006, Rodewald et al. 2010, Lerman et al. 2012, Paker et al. 2013).

Loss et al. (2013) document cat impacts on birds in the United States. Calvert et al. (2013) did a similar study for Canada. Jarvis (2011a, 2011b) reviews estimates of cat abundance and effect in Europe and Australia. Forbush (1916) first brought destructive actions of cats to the attention of ornithologists.

Paker et al. (2013) document slight effects of dogs on bird community diversity. Toxoplasmosis, a disease contracted from cat feces, was a major source of mortality in the endangered Hawaiian crow (Work et al. 2002).

The American Bird Conservancy's program for indoor cats can be found at http://www.abcbirds.org/abcprograms/policy/cats/index.html.

Loss et al. (2014) used a variety of datasets to estimate the number of birds killed in the contiguous United States to window collisions just as they had done for losses to cats. They confirmed Klem's (2008) claim that window collisions were the second leading cause of bird deaths, taking a median value of 599 million birds per year (estimate ranged from 365 million to 988 million). Calvert et al. (2013) calculated losses to collision by birds in Canada. Similar studies have not been done in other countries. The reflective glass in Quebec's downtown is responsible for massive bird mortality (Pouliot 2008).

Klem (2009) presents research findings on the effectiveness of ultraviolet-reflecting glass to reduce avian collisions. Milus (2013) reviews the magnitude of bird losses to building collisions, efforts to reduce it, and tips for improving the safety of buildings.

Strategies to reduce the threat of windows to birds can be found at Chicago Bird Collision Monitors, http://www.birdmonitors.net/, and Flap Canada, http://flap.org.

Additional strategies as well as a source for "bird tape" can be found at the American Bird Conservancy, "Bird Collisions with Glass and Buildings," http://www.abcbirds.org/abcprograms/policy/collisions/glass.html. Ultraviolet-reflective decals can be purchased from Wild Birds Unlimited at http://www.wbu.com/ or WindowAlert at http://windowalert.com.

Efforts and legal mandates to make commercial buildings more bird friendly are discussed at SustainableBusiness.com, http://www.sustainablebusiness.com/index.cfm/go/news.display/id/25063.

Comprehensive resources to reduce the dangers of buildings to birds are available from the Dark Sky Society, "Bird-Safe Building Guidelines," http://www.darkskysociety.org/handouts/birdsafebuildings.pdf, and the American Bird Conservancy, "Bird-Friendly Building Design," http://www.abcbirds.org/newsandreports/BirdFriendlyBuildingDesign.pdf.

Rich and Longcore (2006) provide a comprehensive review of the ecological effects of night lighting.

Estimates of bird mortality resulting from collisions with towers can be found in Longcore et al. (2012, 2013) and Calvert et al. (2013). Manville (2008) suggests how the U.S. Fish and Wildlife Service is working to reduce bird collisions with a variety of structures.

Gehring et al. (2009) conducted experiments that show the benefits of reducing tower collisions by eliminating steady red lights. Poot et al. (2008) find that blue and green lights are less disorienting to birds than are red and white lights. Wiltschko and Wiltschko (2001) demonstrate that red lights interfere with the magnetic compass that migrating birds use to navigate.

Tower operators are keen to eliminate red lights to save money and birds. FCC regulations now allow certain towers to reduce lighting. The steps to do so can be found from Michigan State University's Fewer Lights, Safer Flights, "Toolkit for Operators and Engineers," http://fewerlights.anr.msu.edu/toolkit-for-operators-and-engineers.html.

Principles to reduce the effects of lighting around your home and in your city are presented by Eisenbeis and Hänel (2009).

Lights Out programs are described at the following websites: Chicago Bird Collision Monitors, http://www.birdmonitors.net; Lights Out Indy, http://lightsoutindy.org; Audubon, http://bird-friendly.audubon.org/lightsout; and Flap Canada, http://www.flap.org/lights.php. Such programs are not yet established beyond North America.

The influence of feeders on birds is reviewed by Dunn and Tessaglia-Hymes (1999). Chickadee survival in the presence of feeders was studied by Brittingham and Temple

(1988, 1992a, 1992b). The British Trust for Ornithology's recommendations are reported in Toms (2003).

Suggestions for proper feed, feeders, nest boxes, and their maintenance can be found in Adams (1994), Dunn and Tessaglia-Hymes (1999), and online at Audubon, "Bird Feeding Basics," http://web4.audubon.org/bird/at_home/bird_feeding, and the Cornell Lab of Ornithology, All about Birds, "Feeding Birds," http://www.birds.cornell.edu/AllAboutBirds/faq/master_folder/bird_feeding.

Our study of crow feeders was conducted in 2013 (Marzluff and Miller 2014).

The importance of water, especially in dry climates, to birds was suggested by Adams (1994), Dunn and Tessaglia-Hymes (1999), Walker et al. (2009), and Lepczyk et al. (2012). Rodewald (2012) suggests that the presence of other birds is an important attractor of birds to urban areas.

Snowy owls erupt from their northern climes every four years or so. In the winter of 2013 three were shot at New York's John F. Kennedy airport out of mistaken concern for airplane safety (*Daily News*, "Killing of Snowy Owls at New York's Kennedy Airport Prompts Suit against Federal Agencies by Friends of Animals Advocacy Group," http://www.nydailynews.com/new-york/lawsuit-snowy-owls-article-1.1558491). The snowy owl perched on a cell tower (with blinking red light) in Clearview, Washington, on 8 February 2005.

Keeping predators from becoming habituated to human subsidies is a key to their survival. Strategies to live with carnivores are discussed in depth by Clark et al. (2005), and useful references for homeowners and others are available online: U.S. Forest Service, "Living with Carnivores," http://www.fs.fed.us/outdoors/naturewatch/implementation/Human-Wildlife-Interactions/living-with-carnivores.pdf, and Living with Bears, http://www.livingwithbears.org/livingwithbears.

Bartos et al. (2011) and Mineau and Palmer (2013) have reviewed the hazards of rodent poisons and other household toxins to birds.

Marzluff (2005) and Marzluff and Rodewald (2008) espoused the value of not doing the same thing everywhere (Bunnell 1999) in urban areas. Groffman et al. (2014) document the increasing similarity in the arrangement of land cover and its ecological implication in urban areas across the United States.

Learn about Monroe's Vaux's swifts at Monroe Swift Watch, http://monroeswifts.org. Seideman (2013) describes a similar swift event in Portland, Oregon.

Wheater (2011) documented the diversity of lichens on London's stone walls.

Forman (2008) details the strategies to improve connectivity within urban regions. Information on rails-to-trails programs outside the United States is available from the Rails-to-Trails Conservancy, "A View from the Baltic," http://www.railstotrails .org/news/magazine/webExclusives/2013_Winter_View-From.html; "A View from Down Under," http://www.railstotrails.org/news/magazine/webExclusives/2012_Spring -Summer_View-From.html; "Canadian Spirit, the Trans-Canada Trail," http://www .railstotrails.org/news/magazine/webExclusives/2012_Winter_Third-Feature.html, and "Spain: La Senda Del Oso, the Bear's Path Greenway," http://www.railstotrails .org/resources/documents/magazine/2010_Spring-Summer_Destination.pdf. The im-portance of wetlands in urban areas is further emphasized by Faulkner (2004). Colding (2007) promotes seeking complementary ecological land use during the planning process.

Marzluff and Ewing (2001), Marzluff and Bradley (2003), and Marzluff (2005) offer general guidelines for planners, developers, restoration ecologists, and homeowners wishing to improve the ecological function of urban systems.

Salamander movements around Stanford University are described at Stanford University Habitat Conservation Plan, "California Tiger Salamander," http://hcp .stanford.edu/salamander.html. A general discussion of roads in urban areas is pro-vided by van der Ree (2009). A variety of animal bridges, including in Banff National Park, can be seen at the Wikipedia article "Wildlife Crossing," http://en.wikipedia.org /wiki/Wildlife_crossing. Calvert et al. (2013) document bird mortality during the mow-ing on road edges. Wooded streets may increase bird use and movement (Fernández-Juricic 2000). Forman and Sperling (2011) imagine a world where roads no longer isolate nature. The Central Intelligence Agency's measurement of road length can be found at "The World Factbook," https://www.cia.gov/library/publications/the-world -factbook/fields/2085.html.

Beatley (2011) defines the biophilic city and provides strategies to make cities more "life friendly" and indicators that planners and designers can use to gauge progress to-ward meeting biophilic goals.

Steve Humphrey restores his cloud forest because his dear friend, and noted tropi-cal ecologist, Tom Lovejoy once told him that growing trees was an effective way to ameliorate the climate-altering effects of our carbon dioxide emissions.

Chapter 10. Nature's Tenth Commandment

Jack DeLap, Marcos Garcia, and I hiked with fifteen students into La Serena, Corcovado National Park, on 12 September 2013.

Chape et al. (2005) and Soutullo (2010) summarize statistics for global protected areas. The World Database on Protected Areas is available at http://www.wdpa.org.

Wilson (2002) suggests that tropical wilderness and shallow sea areas that contain around 70 percent of Earth's plant and animal diversity could be saved by a single investment of $30 billion. He reports that in 2000, governments and private sources annually spent $6 billion to sustain biodiversity.

Soulé (1991) and Shafer (1997) provide examples of reserve design that are especially tailored to urban settings.

The inability of reserves to function alone in the conservation of biodiversity is articulated by Mora and Sale (2011). These authors also report on the costs of effective reserves.

Livezey (2009) documents the spread of barred owls to the Pacific Northwest.

Turner et al. (2004) and Miller (2005) summarize environmental amnesia. Dunn et al. (2006) recognized the paradox of global biodiversity protection hinging on humanity's need to connect with urban nature as "the Pigeon Paradox."

Endlicher et al. (2011) provide examples of shrinking cities and their causes and consequences. Plöger (2012) details the loss of manufacturing jobs in Leipzig.

The importance of the city as the place to connect people with nature is emphasized by Marzluff (2002), Miller and Hobbs (2002), and Beatley (2011).

Krasny and Tidball (2012) describe the practice and advantages of "Civic Ecology," including community gardens and their importance to emotional and psychological well-being. Forman (2008) discusses increasing participation in urban agriculture.

Thompson et al. (1999), Florgård (2008), and Dawe (2011) describe the value of urban trees. Luttik (2000), Oleyar et al. (2008), and Farmer et al. (2013) quantify the value of trees and birds to home value.

Kaplan (2011), Matsuoka and Sullivan (2011), Nicholson-Lord (2011), Tilt (2011), and Krasny and Tidball (2012) discuss the health and aesthetic benefits of green neighborhoods.

New York City saves billions of dollars by drawing unfiltered water from reservoirs protected by a series of forested reserves, including the largest in the Catskill Mountains. Maps and details are available from the U.S. Environmental Protection Agency, "Watershed Progress: New York City Watershed Agreement," http://water.epa.gov/type/watersheds/nycityfi.cfm.

I rode with the Indianapolis bird busters on 6 February 2013. Anne Maschmeyer provided cost estimates to me. The issue of equity in confronting ecological issues, generally referred to as "environmental justice," is receiving increasing attention (e.g.,

Kitchen 2013). Van Gelder (1999) articulates the importance of considering equity in the development of sustainable ethics.

Turner et al. (2004), Evans et al. (2005), Miller (2005, 2006), DeStefano (2010), Beatley (2011), and Wells and Lekies (2012) discuss the influence of urban experiences on our ethics and attitudes.

A calendar of birding festivals is provided at Nature Travel Network, "Upcoming Birding and Nature Festivals," http://www.naturetravelnetwork.com/birding-and-nature-festivals-list.

Forman (2008) describes the lack of planning and the need for it as urban regions grow. Beatley (2011) summarizes the planning movement toward sustainability and "biophilic city" design.

Marzluff (2005) and Donnelly and Marzluff (2004a) describe the tipping point for Seattle's birds. Similar thresholds have been reported by With and Crist (1995), Natuhara and Hashimoto (2009), Windmiller et al. (2008), Pennington and Blair (2012), and Yan (2013). Alberti and Marzluff (2004) discuss resilience in urban ecosystems.

Vitousek (1994), Walther et al. (2005), Root and Schneider (2006), and Heller and Zavaleta (2009) review some challenges of climate change to plants and animals.

Dawe (2011) provides estimates of the carbon sequestered by urban street trees.

Jenkins et al. (2003) provide the equations I used to calculate biomass of my trees. Using an average diameter-at-breast-height (dbh) of 48.7 cm and the Douglas-fir equations resulted in a biomass of 1,428 kg for my trees and a total of 516,936 kg for the 362 trees in my forest (half of which, or 258,468 kg, is carbon). By assessing tree diameter in 1980 (40 cm) and 2000 (48 cm), I determined that an average tree puts on 24.7 kg of biomass (12.5 kg of carbon) per year. Therefore, my forest adds 4,525 kg of carbon per year aboveground. Each tree that adds 12.5 kg of carbon uses 45.8 kg of carbon dioxide (1 unit of carbon dioxide = 1 unit of carbon/0.2727).

I used the Terrapass Carbon Footprint Calculator (http://terrapass.com/carbon-footprint-calculator-2) on 11 November 2013 to estimate my household carbon footprint. We drive a Jetta TDI using 20 percent biodiesel and a Subaru Forester a total of ten thousand miles per year. Together they produce 7,693 pounds of carbon dioxide per year. We fly three medium and two long-distance round-trip flights per year that produce 7,750 pounds of carbon dioxide. Our home is heated with a mix of gas and electric, which together produce 12,594 pounds of carbon dioxide. In total, we produce 29,000 pounds of carbon dioxide, which equals 7,908 pounds of carbon (each pound of carbon dioxide is 0.2727 pounds of carbon).

Qian and Follett (2002) and Bandaranayake et al. (2003) estimate carbon budgets of turf.

Adams et al. (2005), DeStefano (2010), and Sterba (2012) discuss the difficulty of managing abundant species in urban areas.

Evans et al. (2005) describe the Smithsonian program to engage citizens in the collection of bird nesting data. The recent book edited by Dickinson and Bonney (2012) provides a broad overview of citizen science, including tracing its roots to President Jefferson.

Busch (2013) chronicles the experience of being a citizen scientist and has a wonderful appendix of citizen science opportunities. More projects can be found at the Citizen-Sci blog, http://blogs.plos.org/citizensci.

Holly Parsons introduced me to the powerful owl project during her talk at the 2011 Urban Bird Conservation workshop held at the International Congress for Conservation Biology (Fergus et al. 2012).

Fischer et al. (2012) suggest how to change human behavior toward more sustainable actions.

References

Able, K. P., and J. R. Belthoff. 1998. Rapid "evolution" of migratory behavior in the introduced house finch of eastern North America. *Proceedings of the Royal Society of London B* 265:2063–2071.

Adams, L. W. 1994. *Urban Wildlife Habitats: A Landscape Perspective.* Minneapolis: University of Minnesota Press.

Adams, L. W., L. W. VanDruff, and M. Luniak. 2005. Managing urban habitats and wildlife. Pages 714–739 in *Techniques for Wildlife Investigation and Management*, edited by C. Braun. Bethesda, MD: Wildlife Society.

Akçakaya, H. R. 2002. *RAMAS GIS: Linking Spatial Data with Population Viability Analysis* (version 4.0). Setauket, NY: Applied Biomathematics.

Alberti, M. 2008. *Advances in Urban Ecology.* New York: Springer.

Alberti, M., and J. M. Marzluff. 2004. Ecological resilience in urban ecosystems: linking urban patterns to human and ecological functions. *Urban Ecosystems* 7:241–265.

Aldrich, J. W., and R. W. Coffin. 1980. Breeding bird populations from forest to suburbia after thirty-seven years. *American Birds* 34:3–7.

Andrews, K. M., J. W. Gibbons, and D. M. Jochimsen. 2008. Roads as catalysts of urbanization: snakes on roads face differential impacts due to inter- and intraspecific ecological attributes. Pages 145–153 in *Urban Herpetology*, edited by J. C. Mitchell, R. E. Jung Brown, and B. Bartholomew. Salt Lake City: Society for the Study of Amphibians and Reptiles.

Angel, S., J. Parent, D. L. Civco, A. Blei, and D. Potere. 2011. The dimensions of global urban expansion: estimates and projections for all countries, 2000–2050. *Progress in Planning* 75:53–107.

Arendt, R. G. 1996. *Conservation Design for Subdivisions: A Practical Guide to Creating Open Space Networks.* Washington, DC: Island Press.

Arroyo-Solís, A., J. M. Castillo, E. Figueroa, J. L. López-Sánchez, and H. Slabbekoorn. 2013. Experimental evidence for an impact of anthropogenic noise on dawn chorus timing in urban birds. *Journal of Avian Biology* 44:288–296.

Atwell, J. W., G. C. Cardoso, D. J. Whittaker, S. Campbell-Nelson, K. W. Robertson, and E. D. Ketterson. 2012. Boldness behavior and stress physiology in a novel urban environment suggest rapid correlated evolutionary adaptation. *Behavioral Ecology* 23:960–969.

Baillie, J., and B. Groombridge, eds. 1996. *IUCN Red List of Threatened Animals.* Gland, Switzerland: IUCN.

Baker, A. J. 1980. Morphometric differentiation in New Zealand populations of the house sparrow (*Passer domesticus*). *Evolution* 34:638–653.

Baker, A. J., and A. Moeed. 1979. Evolution in the introduced New Zealand populations of the common myna, *Acridotheres tristis* (Aves: Sturnidae). *Canadian Journal of Zoology* 57:570–584.

Baker, P. J., A. J. Bentley, R. J. Ansell, and S. Harris. 2005. Impact of predation by domestic cats *Felis catus* in an urban area. *Mammal Review* 35:302–312.

Bandaranayake, W., Y. L. Qian, W. J. Parton, D. S. Ojima, and R. F. Follett. 2003. Estimation of soil organic carbon changes in turfgrass systems using the CENTURY model. *Agronomy Journal* 95:558–563.

Banks, R. C. 1997. In Memoriam: John Warren Aldrich, 1906–1995. *Auk* 114:748–751.

Barratt, D. G. 1998. Predation by house cats, *Felis catus* (L.), in Canberra, Australia. II. Factors affecting the amount of prey caught and estimates of the impact on wildlife. *Wildlife Research* 25:475–487.

Bartos, M., S. Dao, D. Douk, S. Falzone, E. Gumerlock, S. Hoekstra, K. Kelly-Relf, D. Mori, C. Tang, C. Vasquez, J. Ward, S. Young, A. T. Morzillo, S. P. D. Riley, and T. Longcore. 2011. Use of anticoagulant rodenticides in single-family neighborhoods along an urban-wildland interface in California. *Cities and the Environment* 4(1):12. Available at http://digitalcommons.lmu.edu/cate/vol4/iss1/12.

Beahop, S., W. Fiedler, R. W. Furness, S. C. Votier, S. Waldron, J. Newton, G. J. Bowen, P. Berthold, and K. Farnsworth. 2005. Assortative mating as a mechanism for rapid evolution of a migratory divide. *Science* 310:502–504.

Beatley, T. 2011. *Biophilic Cities, Integrating Nature into Urban Design and Planning.* Washington, DC: Island Press.

Bell, C. P., S. W. Baker, N. G. Parkes, M. Brook, and D. E. Chamberlain. 2010. The role of the Eurasian sparrowhawk (*Accipiter nisus*) in the decline of the house sparrow (*Passer domesticus*) in Britain. *Auk* 127:411–420.

Berkowitz, A. R., C. H. Nilon, and K. S. Hollweg, eds. 2003. *Understanding Urban Ecosystems*. New York: Springer.

Berry, B. J. L. 1990. Urbanization. Pages 103–119 in *The Earth as Transformed by Human Action*, edited by B. L. Turner II, W. C. Clark, R. W. Kates, J. F. Richards, J. T. Mathews, and W. B. Meyer. Cambridge: Cambridge University Press.

Berthold, P., A. J. Helbig, G. Mohr, and U. Querner. 1992. Rapid microevolution of migratory behavior in a wild bird species. *Nature* 360:668–670.

Bird, D., D. Varland, and J. Negro, eds. 1996. *Raptors in Human Landscapes: Adaptations to Built and Cultivated Environments*. London: Academic Press.

Bishop, J. A., and L. M. Cook. 1980. Industrial melanism and the urban environment. *Advances in Ecological Research* 11:373–404.

Bittner, M. 2004. *The Wild Parrots of Telegraph Hill: A Love Story . . . with Wings*. New York: Harmony Books.

Blair, R. B. 1996. Land use and avian species diversity along an urban gradient. *Ecological Applications* 6:506–519.

——. 2001. Birds and butterflies along urban gradients in two ecoregions of the United States: is urbanization creating a homogeneous fauna? Pages 33–56 in *Biotic Homogenization*, edited by J. L. Lockwood and M. L. McKinney. New York: Kluwer Academic/Plenum.

Blaustein, A. R., and P. T. J. Johnson. 2003. The complexity of deformed amphibians. *Frontiers in Ecology and the Environment* 1:87–94.

Bled, F., A. J. Royle, and E. Cam. 2011. Hierarchical modeling of an invasive spread: the Eurasian collared-dove *Streptopelia decaocto* in the United States. *Ecological Applications* 21:290–302.

Blem, C. R. 1974. Geographic variation of thermal conductance in the house sparrow *Passer domesticus*. *Comparative Biochemistry and Physiology* 47A:101–108.

Blewett, C. M., and J. M. Marzluff. 2005. Effects of urban sprawl on snags and the abundance and productivity of cavity-nesting birds. *Condor* 107:678–693.

Boal, C. W., and R. W. Mannan. 1998. Nest-site selection by Cooper's Hawks in an urban landscape. *Journal of Wildlife Management* 63:77–84.

Bock, C. E., Z. F. Jones, and J. H Bock. 2008. The oasis effect: response of birds to exurban development in a southwestern savanna. *Ecological Applications* 18:1093–1106.

Bonnington, C., K. J. Gaston, and K. L. Evans. 2013. Fearing the feline: domestic cats reduce avian fecundity through trait-mediated indirect effects that increase nest predation by other species. *Journal of Applied Ecology* 50:15–24.

Bonter, D. N., B. Zuckerberg, and J. L. Dickinson. 2010. Invasive birds in a novel landscape: habitat associations and effects on established species. *Ecography* 33:494–502.

Bormann, F. H., D. Balmori, and G. T. Geballe. 2001. *Redesigning the American Lawn: A Search for Environmental Harmony*. 2nd ed. New Haven, CT: Yale University Press.

Brisbin, I. L., R. A. Kennamer, E. L. Peters, and D. J. Karapatakis. 2008. A long-term study of eastern box turtles (*Terrapene c. carolina*) in a suburban neighborhood: survival characteristics and interactions with humans and conspecifics. Pages 373–385 in *Urban Herpetology*, edited by J. C. Mitchell, R. E. Jung Brown, and B. Bartholomew. Salt Lake City: Society for the Study of Amphibians and Reptiles.

Brittingham, M. C., and S. A. Temple. 1988. Impacts of supplemental feeding on survival rates of black-capped chickadees. *Ecology* 69:581–589.

———. 1992a. Does winter bird feeding promote dependency? *Journal of Field Ornithology* 63:190–194.

———. 1992b. Use of winter bird feeders by black-capped chickadees. *Journal of Wildlife Management* 56:103–110.

Brown, C. R., and M. B. Brown. 2013. Where has all the road kill gone? *Current Biology* 23:233–234.

Bruce, J. P., J. S. Quinn, S. A. Sloane, and B. N. White. 1996. DNA fingerprinting reveals monogamy in the Bushtit, a cooperatively breeding species. *Auk* 113:511–516.

Bruno, J. F., J. J. Stachowicz, and M. D. Bertness. 2003. Inclusion of facilitation into ecological theory. *Trends in Ecology and Evolution* 18:119–125.

Bunnell, F. L. 1999. What habitat is an island? Pages 1–31 in *Forest Fragmentation: Wildlife and Management Implications*, edited by J. A. Rochelle, L. A. Lehmann, and J. Wisniewski. Boston: Brill.

Burdett, C. L., K. R. Crooks, D. M. Theobald, K. R. Wilson, E. E. Boyston, L. M. Lyren, R. N. Fisher, T. W. Vickers, S. A. Morrison, and W. M. Boyce. 2010. Interfacing models of wildlife habitat and human development to predict the future distribution of puma habitat. *Ecosphere* 1(1):4. Available at http://www.esajournals.org/doi/full/10.1890/ES10-00005.1

Burroughs, J. 1871. *Wake Robin*. Boston: Houghton Mifflin.

Busch, A. 2013. *The Incidental Steward: Reflections on Citizen Science*. New Haven, CT: Yale University Press.

Calvert, A. M., C. A. Bishop, R. D. Elliot, E. A. Krebs, T. M. Kydd, C. S. Machtans, and G. J. Robertson. 2013. A synthesis of human-related avian mortality in Canada.

Avian Conservation and Ecology 8(2):11. Available at http://dx.doi.org/10.5751/ACE -00581-080211.

Cam, E., J. D. Nichols, J. R. Sauer, J. E. Hines, and C. H. Flather. 2000. Relative species richness and community completeness: birds and urbanization in the mid-Atlantic States. *Ecological Applications* 10:1196–1210.

Catterall, C. P. 2009. Responses of faunal assemblages to urbanization: global research paradigms and an avian case study. Pages 129–155 in *Ecology of Cities and Towns: A Comparative Approach*, edited by M. J. McDonnell, A. K. Hahs, and J. H. Breuste. Cambridge: Cambridge University Press.

Catterall, C. P., M. B. Kingston, K. Park, and S. Sewell. 1998. Deforestation, urbanization and seasonality: interacting effects on a regional bird assemblage. *Biological Conservation* 84:65–81.

Cavia, R., G. R. Cueto, and O. V. Suárez. 2009. Changes in rodent communities according to the landscape structure of an urban ecosystem. *Landscape and Urban Planning* 90:11–19.

Chace, J. F., and J. J. Walsh. 2006. Urban effects on native avifauna: a review. *Landscape and Urban Planning* 74:46–69.

Chamberlain, D. E., A. R. Cannon, M. P. Toms, D. I. Leech, B. J. Hatchwell, and K. J. Gaston. 2009. Avian productivity in urban landscapes: a review and meta-analysis. *Ibis* 151:1–18.

Chape, S., J. Harrison, M. Spalding, and I. Lysenko. 2005. Measuring the extent and effectiveness of protected areas as an indicator for meeting global biodiversity targets. *Philosophical Transactions of the Royal Society B* 360:443–455.

Chapin, F. S., III, E. S. Zavaleta, V. T. Eviner, R. L. Naylor, P. M. Vitousek, H. L. Reynolds, D. U. Hooper, S. Lavorel, O. E. Sala, S. E. Hobbie, M. C. Mack, and S. Diaz. 2000. Consequences of changing biodiversity. *Nature* 305:234–242.

Chapman, M. G., D. Blockley, J. People, and B. Clynick. 2009. Effect of urban structures on diversity of marine species. Pages 156–176 in *Ecology of Cities and Towns: A Comparative Approach*, edited by M. J. McDonnell, A. K. Hahs, and J. H. Breuste. Cambridge: Cambridge University Press.

Chapman, M. G., and A. J. Underwood. 2009. Comparative effects of urbanization in marine and terrestrial habitats. Pages 51–70 in *Ecology of Cities and Towns: A Comparative Approach*, edited by M. J. McDonnell, A. K. Hahs, and J. H. Breuste. Cambridge: Cambridge University Press.

Chiappe, L. M., and G. J. Dyke. 2001. The Mesozoic radiation of birds. *Annual Review of Ecology and Systematics* 33:91–124.

Christman, B. J., and A. A. Dhondt. 1997. Nest predation in black-capped chickadees: how safe are cavity nests? *Auk* 114:769–773.

Clark, P. J., M. J. Reed, B. G. Tavernia, B. S. Windmiller, and J. V. Regosin. 2008. Urbanization effects on spotted salamander and wood frog presence and abundance. Pages 67–75 in *Urban Herpetology*, edited by J. C. Mitchell, R. E. Jung Brown, and B. Bartholomew. Salt Lake City: Society for the Study of Amphibians and Reptiles.

Clark, T., M. Rutherford, and D. Casey, eds. 2005. *Coexisting with Large Carnivores: Lessons from Greater Yellowstone*. Washington, DC: Island Press.

Clergeau, P., S. Croci, J. Jokimäki, M.-L. Kaisanlahti-Jokimäki, and M. Dinetti. 2006. Avifauna homogenization by urbanization: analysis at different European latitudes. *Biological Conservation* 127:336–344.

Clews, B. 2006. *Birds in a Village: A Century On*. Hampshire, UK: Wildguides.

Clucas, B., and J. M. Marzluff. 2011. Coupled relationships between humans and animals in urban areas. Pages 135–147 in *Handbook of Urban Ecology*, edited by J. Niemela. Oxford: Oxford University Press.

———. 2012. Attitudes and actions toward birds in urban areas: human cultural differences influence bird behavior. *Auk* 129:8–16.

Clucas, B., J. M. Marzluff, S. Kübler, and P. Meffert. 2011. New directions in urban avian ecology: reciprocal connections between birds and humans in cities. Pages 167–196 in *Perspectives in Urban Ecology*, edited by W. Endlicher, P. Hostert, I. Kowarik, E. Kulke, J. Lossau, J. Marzluff, E. van der Meer, H. Mieg, G. Nützmann, M. Schulz, and G. Wessolek. New York: Springer.

Clucas, B., J. M. Marzluff, D. Mackovjak, and I. Palmquist. 2013. Do American crows pay attention to human gaze and facial expression? *Ethology* 119:296–302.

Cockle, K. L., K. Martin, and G. Robledo. 2012. Linking fungi, trees, and hole-using birds in a neotropical tree-cavity network: pathways of cavity production and implications for conservation. *Forest Ecology and Management* 264:210–219.

Colding, J. 2007. "Ecological land-use complementation" for building resilience in urban ecosystems. *Landscape and Urban Planning* 81:46–55.

Colding, J., and C. Folke. 2009. The role of golf courses in biodiversity conservation and ecosystem management. *Ecosystems* 12:191–206.

Connell, J. H. 1980. Diversity and the coevolution of competitors, or the ghost of competition past. *Oikos* 35:131–138.

Cook, L. M., G. S. Mani, and M. E. Varley. 1986. Postindustrial melanism in the peppered moth. *Science* 231:611–613.

Cornell, K. L., C. R. Kight, R. B. Burdge, A. R. Gunderson, J. K. Hubbard, A. K. Jackson, J. E. LeClerc, M. L. Pitts, J. P. Swaddle, and D. A. Cristol. 2011. Reproductive success of eastern bluebirds (*Sialia sialis*) on suburban golf courses. *Auk* 128:577–586.

Coyne, J. A. 2009. *Why Evolution Is True*. New York: Viking Penguin.

Cox, W. 2012. World urban areas population and density: A 2012 update. 2 March 2014. Available at http://www.newgeography.com/content/002808-world-urban-areas -population-and-density-a-2012-update.

Cramp, S. 1980. Changes in the breeding birds of inner London since 1900. *Proceedings of the International Ornithological Congress* 17:1316–1320.

Cramp, S., and A. D. Tomlins. 1966. The birds of inner London, 1951–1965. *British Birds* 59:209–233.

Cristol, D. A., and A. D. Rodewald. 2005. Introduction: can golf courses play a role in bird conservation? *Wildlife Society Bulletin* 33:407–410.

Czech, B., and P. R. Krausman. 1997. Distribution and causation of species endangerment in the United States. *Science* 277:1116–1117.

Czech, B., P. R. Krausman, and P. K. Devers. 2000. Economic associations among causes of species endangerment in the United States. *BioScience* 50:593–601.

Daniels, T. 1999. *When City and Country Collide*. Washington, DC: Island Press.

Darwin, C. R. 1859. *On the Origin of Species by Means of Natural Selection, or the Preservation of Favoured Races in the Struggle for Life*. London: John Murray.

Davis, A., C. E. Taylor, and R. E. Major. 2012. Seasonal abundance and habitat use of Australian parrots in an urbanized landscape. *Landscape and Urban Planning* 106:191–198.

Dawe, G. F. M. 2011. Street trees and the urban environment. Pages 424–449 in *The Routledge Handbook of Urban Ecology*, edited by I. Douglas, D. Goode, M. C. Houck, and R. Wang. London: Routledge.

Dawson, W. L., with J. H. Bowles; drawings by Allan Brooks. 1909. *The Birds of Washington*. Seattle: Occidental.

DeStefano, S. 2010. *Coyote at the Kitchen Door: Living with Wildlife in Suburbia*. Cambridge, MA: Harvard University Press.

Devictor, V., R. Julliard, D. Couvet, A. Lee, and F. Jiguet. 2007. Functional homogenization effect of urbanization on bird communities. *Conservation Biology* 21:741–751.

Dickinson, J. L., and R. Bonney, eds. 2012. *Citizen Science: Public Participation in Environmental Research*. Ithaca, NY: Cornell University Press.

Diller, L. V. 2014. To shoot or not to shoot. *Wildlife Society News*. January 17. Available at http://news.wildlife.org/featured/to-shoot-or-not-to-shoot.

Dinetti, M. 2007. *I Passeri Passer spp*. Nelle aree urbane e nel territorio in Italia. Distribuzione, densità e status di conservazione: una review. *Ecologia Urbana* 19:11–42.

Dobkin, D. S., A. C. Rich, J. A. Pretare, and W. H. Pyle. 1995. Nest-site relationships among cavity-nesting birds of riparian and snowpocket aspen woodlands in the northwestern Great Basin. *Condor* 97:694–707.

Donnelly, R., and J. M. Marzluff. 2004a. Importance of reserve size and landscape context to urban bird conservation. *Conservation Biology* 18:733–745.

———. 2006. Relative importance of habitat quantity, structure, and spatial pattern to birds in urbanizing environments. *Urban Ecosystems* 9:99–117.

Douglas, I., D. Goode, M. C. Houck, and R. Wang, eds. 2011. *The Routledge Handbook of Urban Ecology*. London: Routledge.

Drever, M. C., K. E. H. Aitken, A. R. Norris, and K. Martin. 2008. Woodpeckers as reliable indicators of bird richness, forest health and harvest. *Biological Conservation* 141:624–634.

Dugger, K. M., R. G. Anthony, and L. S. Andrews. 2011. Transient dynamics of invasive competition: barred owls, spotted owls, habitat, and the demons of competition present. *Ecological Applications* 21:2459–2468.

Dunn, E. H., and D. L. Tessaglia-Hymes. 1999. *Birds at Your Feeder: A Guide to Feeding Habits, Behavior, Distribution, and Abundance*. New York: W. W. Norton.

Dunn, R. R., M. C. Gavin, M. C. Sanchez, and J. N. Solomon. 2006. The pigeon paradox: dependence of global conservation on urban nature. *Conservation Biology* 20:1814–1816.

Dwyer, J. F., D. J. Nowak, M. H. Noble, and S. M. Sisinni. 2000. Connecting people with ecosystems in the 21st century: an assessment of our nation's urban forests. General Technical Report PNW-GTR-490. Portland, OR: U.S. Department of Agriculture, Forest Service.

Egan, R. S., and P. C. Paton. 2008. Multiple scale habitat characteristics of pond-breeding amphibians across a rural-urban gradient. Pages 53–65 in *Urban Herpetology*, edited by J. C. Mitchell, R. E. Jung Brown, and B. Bartholomew. Salt Lake City: Society for the Study of Amphibians and Reptiles.

Eisenbeis, G., and A. Hänel. 2009. Light pollution and the impact of artificial night lighting on insects. Pages 243–263 in *Ecology of Cities and Towns: A Comparative*

Approach, edited by M. J. McDonnell, A. K. Hahs, and J. H. Breuste. Cambridge: Cambridge University Press.

Endlicher, W. 2011. Introduction: from urban nature studies to ecosystem services. Pages 1–13 in *Perspectives in Urban Ecology*, edited by W. Endlicher, P. Hostert, I. Kowarik, E. Kulke, J. Lossau, J. Marzluff, E. van der Meer, H. Mieg, G. Nützmann, M. Schulz, and G. Wessolek. New York: Springer.

Endlicher, W., P. Hostert, I. Kowarik, E. Kulke, J. Lossau, J. Marzluff, E. van der Meer, H. Mieg, G. Nutzmann, M. Schulz, and G. Wessolek, eds. 2011. *Perspectives in Urban Ecology: Studies of Ecosystems and Interactions between Humans and Nature in the Metropolis of Berlin*. Berlin: Springer.

Er, K. G. H., J. L. Innes, K. Martin, and B. Klinkenberg. 2005. Forest loss with urbanization predicts bird extirpations in Vancouver. *Biological Conservation* 126:410–419.

Ervin, S. 1975. Iris coloration in young bushtits. *Condor* 77:90–107.

Evans, C., E. Abrams, R. Reitsma, K. Roux, L. Salmonsen, and P. P. Marra. 2005. The Neighborhood Nestwatch program: participant outcomes of a citizen-science ecological research project. *Conservation Biology* 19:589–594.

Ewing, R. H. 1994. Characteristics, causes, and effects of sprawl: a literature review. *Environmental and Urban Issues* 21:1–15.

Faeth, S. H., P. S. Warren, E. Shochat, and W. A. Marussich. 2005. Trophic dynamics in urban communities. *BioScience* 55:399–407.

Farmer, M. C., M. C. Wallace, and M. Shiroya. 2013. Bird diversity indicates ecological value in urban home prices. *Urban Ecosystems* 16:131–144.

Farwell, L. S., and J. M. Marzluff. 2012. A new bully on the block: does urbanization promote Bewick's wren (*Thryomanes bewickii*) aggressive exclusion of Pacific wrens (*Troglodytes pacificus*)? *Biological Conservation* 161:128–141.

Faulkner, S. 2004. Urbanization impacts on the structure and function of forested wetlands. *Urban Ecosystems* 7:89–106.

Fedriani, J. M., T. K. Fuller, and R. M. Sauvajot. 2001. Does anthropogenic food enhance densities of omnivorous mammals? An example with coyotes in southern California. *Ecography* 24:325–331.

Ferebe, K. B., and P. E. P. Henry. 2008. Movements and distribution of *Terrapene carolina* in a large urban area, Rock Creek National Park, Washington, DC. Pages 365–372 in *Urban Herpetology*, edited by J. C. Mitchell, R. E. Jung Brown, and B. Bartholomew. Salt Lake City: Society for the Study of Amphibians and Reptiles.

Fergus, R., R. Kwak, and J. Louwe Kooijmans. 2012. Urban bird conservation: for birds and people, a report of a workshop held at the Society for Conservation Biology,

25th International Congress for Conservation Biology, Auckland, New Zealand, 5 December 2011.

Fernández-Juricic, E. 2000. Avifaunal use of wooded streets in an urban landscape. *Conservation Biology* 14:513–521.

Fischer, J., R. Dyball, I. Fazey, C. Gross, S. Dovers, P. R. Ehrlich, R. J. Brulle, C. Christensen, and R. J. Borden. 2012. Human behavior and sustainability. *Frontiers in Ecology and the Environment* 10:153–160.

Florgård, C. 2008. Preservation of original natural vegetation in urban areas: an overview. Pages 380–398 in *Ecology of Cities and Towns: A Comparative Approach*, edited by M. J. McDonnell, A. K. Hahs, and J. H. Breuste. Cambridge: Cambridge University Press.

Forbush, E. H. 1916. *The Domestic Cat*. Economic Biology Bulletin No. 2. Boston: Commonwealth of Massachusetts, State Board of Agriculture.

Forman, R. T. T. 2008. *Urban Regions, Ecology and Planning beyond the City*. Cambridge: Cambridge University Press.

Forman, R. T. T., and D. Sperling. 2011. The future of roads: no driving, no emissions, nature reconnected. *Solutions* 2:10–23.

Fox, S-J. C., and P. A. R. Hockey. 2007. Impacts of a South African coastal golf estate on shrubland bird communities. *South African Journal of Science* 103:27–34.

Francis, T. B., and D. E. Schindler. 2006. Degradation of littoral habitats by residential development: woody debris in lakes of the Pacific Northwest and Midwest, United States. *Ambio* 35:274–280.

———. 2009. Shoreline urbanization reduces terrestrial insect subsidies to fishes in North American lakes. *Oikos* 118:1872–1882.

Friesen, L. E., P. F. J. Eagles, and R. J. Mackay. 1995. Effects of residential development on forest-dwelling neotropical migrant songbirds. *Conservation Biology* 9:1408–1414.

Fry, J., G. Xian, S. Jin, J. Dewitz, C. Homer, L. Yang, C. Barnes, N. Herold, and J. Wickham. 2011. Completion of the 2006 National Land Cover Database for the Conterminous United States. *Photogrammetric Engineering and Remote Sensing* 77:858–864.

Fulgione, D., G. Procaccini, and M. Milone. 2000. Urbanization and the genetic structure of *Passer italiae* (Vieillot 1817) populations in the south of Italy. *Ethology, Ecology and Evolution* 12:123–130.

Fuller, R. A., K. N. Irvine, Z. G. Davies, P. R. Armsworth, and K. J. Gaston. 2012. Interactions between people and birds in urban landscapes. *Studies in Avian Biology* 45:249–266.

Gaston, K. J., and S. Gaston. 2011. Urban gardens and biodiversity. Pages 450–458 in *The Routledge Handbook of Urban Ecology*, edited by I. Douglas, D. Goode, M. Houck, and R. Wang, London: Routledge.

Gehlbach, F. R. 1998. Population and environmental features that promote adaptation to urban ecosystems: the case of Eastern screech-owls (*Otus asio*) in Texas. Pages 1809–1813 in *Proceedings of the XIX Congressus Internationalis Ornithologici*. Ottawa, Canada: National Museum of Natural Science.

———. 2005. Native Texas avifauna altered by suburban entrapment and method for easily assessing natural avifaunal value. *Bulletin of the Texas Ornithological Society* 38:35–47.

Gehring, J., P. Kerlinger, and A. M. Manville II. 2009. Communication towers, lights, and birds: successful methods of reducing the frequency of avian collisions. *Ecological Applications* 19:505–514.

Gering, J. C., and R. B. Blair. 1999. Predation on artificial bird nests along an urban gradient: predatory risk or relaxation in urban environments? *Ecography* 22:532–541.

Germaine, S. S., and B. F. Wakeling. 2001. Lizard species distributions and habitat occupation along an urban gradient in Tucson, Arizona, USA. *Biological Conservation* 97:229–237.

Gibson, M. 2000. Hamoukar, early city in northeastern Syria. Oriental Institute of the University of Chicago, *News and Notes* No. 166.

Giles, J. C., G. Kuchling, and J. A. Davis. 2008. Populations of the snake-necked turtle *Chelodina oblonga* in three suburban lakes of Perth, Western Australia. Pages 275–283 in *Urban Herpetology*, edited by J. C. Mitchell, R. E. Jung Brown, and B. Bartholomew. Salt Lake City: Society for the Study of Amphibians and Reptiles.

Gomulkiewicz, R., and R. D. Holt. 1995. When does evolution by natural selection prevent extinction? *Evolution* 49:201–207.

Goszczyński, J., P. Jabłoński, G. Lesiński, and J. Romanowski. 1993. Variation in diet of Tawny Owl *Strix aluco* L. along an urbanization gradient. *ACTA Ornithologica* 27:113–123.

Grant, P. R., and B. R. Grant. 2011. *How and Why Species Multiply*. Princeton, NJ: Princeton University Press.

Greenberg, R., and P. P. Marra. 2005. *Birds of Two Worlds: The Ecology and Evolution of Migration*. Baltimore: Johns Hopkins University Press.

Groffman, P. M., J. Cavender-Bares, N. D. Bettez, J. M. Grove, S. J. Hall, J. B. Heffernan, S. E. Hobbie, K. L. Larson, J. L. Morse, C. Neill, K. Nelson, J. O'Neil-Dunne, L. Ogden, D. E. Pataki, C. Polsky, R. R. Chowdhury, and M. K. Steele. 2014.

Ecological homogenization of urban USA. *Frontiers in Ecology and the Environment* 12:74–81.

Grubb, T. C., Jr., and D. A. Cimprich. 1990. Supplementary food improves the nutritional condition of wintering woodland birds: evidence from ptilochronology. *Ornis Scandinavica* 21:277–281.

Hacker, S. D., and S. D. Gaines. 1997. Some implications of direct positive interactions for community species diversity. *Ecology* 78:1990–2003.

Hagood, S., and M. J. Bartles. 2008. Use of existing culverts by eastern box turtles (*Terrapene c. carolina*) to safely navigate roads. Pages 169–170 in *Urban Herpetology*, edited by J. C. Mitchell, R. E. Jung Brown, and B. Bartholomew. Salt Lake City: Society for the Study of Amphibians and Reptiles.

Hansen, A. J., R. L. Knight, J. M. Marzluff, S. Powell, K. Brown, P. Hernandez, and K. Jones. 2005. Effects of exurban development on biodiversity: patterns, mechanisms, research needs. *Ecological Applications* 15:1893–1905.

Harestad, A. S., and D. G. Keisker. 1989. Nest tree use by primary cavity-nesting birds in south central British Columbia. *Canadian Journal of Zoology* 67:1067–1073.

Haugen, T. O., and L. A. Vøllestad. 2001. A century of life-history evolution in grayling. *Genetica* 112–113:475–491.

Hedblom, M., and B. Söderström. 2012. Effects of urban matrix on reproductive performance of Great Tit (*Parus major*) in urban woodlands. *Urban Ecosystems* 15:167–180.

Heller, N. E., and E. S. Zavaleta. 2009. Biodiversity management in the face of climate change: a review of 22 years of recommendations. *Biological Conservation* 142:14–32.

Hendry, A. P., and M. T. Kinnison. 1999. Perspective. The pace of modern life: measuring rates of contemporary microevolution. *Evolution* 53:1637–1653.

———. 2001. An introduction to microevolution: rate, pattern, process. *Genetica* 112–113:1–8.

Hermansen, J. S., S. A. Saether, T. O. Elgvin, T. Borge, E. Hjelle, and G.-P. Saetre. 2011. Hybrid speciation in sparrows I: phenotypic intermediacy, genetic admixture and barriers to gene flow. *Molecular Ecology* 20:3812–3822.

Hetmański, T., and A. Jarosiewicz. 2008. Plumage polymorphism and breeding parameters of various feral pigeon (*Columba livia* GM) morphs in urban area (Gdańsk, North Poland). *Polish Journal of Ecology* 56:683–691.

Hochachka, W. M., and A. A. Dhondt. 2000. Density-dependent decline of host abundance resulting from a new infectious disease. *Proceedings of the National Academy of Sciences (USA)* 97:5303–5306.

Hochuli, D. F., F. J. Christie, and B. Lomov. 2009. Invertebrate biodiversity in urban landscapes: assessing remnant habitat and its restoration. Pages 215–232 in *Ecology of Cities and Towns: A Comparative Approach*, edited by M. J. McDonnell, A. K. Hahs, and J. H. Breuste. Cambridge: Cambridge University Press.

Hodgkison, S. C., J.-M. Hero, and J. Warnken. 2007a. The conservation value of suburban golf courses in a rapidly urbanizing region of Australia. *Landscape and Urban Planning* 79:323–337.

———. 2007b. The efficacy of small-scale conservation efforts, as assessed on Australian golf courses. *Biological Conservation* 136:576–586.

Hoffmann, C. O., and J. L. Gottschang. 1977. Numbers, distribution, and movements of a raccoon population in a suburban residential community. *Journal of Mammalogy* 58:623–636.

Horn, D. J. 1999. The species richness of birds visiting a yard is influenced by the feeders/seeds present. *Journal of the Iowa Academy of Sciences* 106:21–25.

Horn, H. S. 1968. The adaptive significance of colonial nesting in the Brewer's blackbird (*Euphagus cyanocephalus*). *Ecology* 49:682–694.

Hostetler, M., and D. Drake. 2009. Conservation subdivisions: a wildlife perspective. *Landscape and Urban Planning* 90:95–101.

Housing Assistance Council. 2011. Estimating Potential Changes to USDA-RD's Eligible Area Designations. Rural Housing Research Note. Washington, DC: Housing Assistance Council.

Hudson, M.-A. R., and D. M. Bird. 2009. Recommendations for design and management of golf courses and green spaces based on surveys of breeding bird communities in Montreal. *Landscape and Urban Planning* 92:335–346.

Ingold, D. J. 1989. Nesting phenology and competition for nest sites among red-headed and red-bellied woodpeckers and European starlings. *Auk* 106:209–217.

———. 1994. Influence of nest-site competition between European starlings and woodpeckers. *Wilson Bulletin* 106:227–241.

———. 1996. Delayed nesting decreases reproductive success in northern flickers: implications for competition with European starlings. *Journal of Field Ornithology* 67:321–326.

Ishchenko, V. G., and J. C. Mitchell. 2008. Urban herpetology in Russia and adjacent territories. Pages 405–421 in *Urban Herpetology*, edited by J. C. Mitchell, R. E. Jung Brown, and B. Bartholomew. Salt Lake City: Society for the Study of Amphibians and Reptiles.

Jackson, A. K., J. P. Froneberger, and D. A. Cristol. 2011. Postfledging survival of eastern bluebirds in an urbanized landscape. *Journal of Wildlife Management* 75:1082–1093.

———. 2012. Habitat near nest boxes correlated with fate of eastern bluebird fledglings in an urban landscape. *Urban Ecosystems* 16:367–376.

Jarvis, P. J. 2011a. Urban animal ecology. Pages 352–360 in *The Routledge Handbook of Urban Ecology*, edited by I. Douglas, D. Goode, M. C. Houck, and R. Wang. London: Routledge.

———. 2011b. Feral animals in the urban environment. Pages 361–369 in *The Routledge Handbook of Urban Ecology*, edited by I. Douglas, D. Goode, M. C. Houck, and R. Wang. London: Routledge.

Jenkins, J. C., D. C. Chojnacky, L. S. Heath, and R. A. Birdsey. 2003. National-scale biomass estimators for United States tree species. *Forest Science* 49:12–35.

Johnson, C., P. J. Trubl, and L. S. Miles. 2012. Black widows in an urban desert: city-living compromises spider fecundity and egg investment despite urban prey abundance. *American Midland Naturalist* 168:333–340.

Johnston, R. F. 1969. Taxonomy of house sparrows and their allies in the Mediterranean basin. *Condor* 71:129–139.

Johnston, R. F., and R. K. Selander. 1964. House sparrows: rapid evolution of races in North America. *Science* 144:548–550.

———. 1971. Evolution in the house sparrow II. Adaptive differentiation in North American populations. *Evolution* 25:1–28.

Johnston, R. F., D. M. Niles, and S. A. Rohwer. 1972. Hermon Bumpus and natural selection in the house sparrow *Passer domesticus*. *Evolution* 26:20–31.

Jokimäki, J., J. Suhonen, K. Inki, and S. Jokinen. 1996. Biogeographical comparison of winter bird assemblages in urban environments in Finland. *Journal of Biogeography* 23:379–386.

Jones, S. G., D. H. Gordon, G. M. Phillips, and B. R. D. Richardson. 2005. Avian community response to a golf-course landscape unit gradient. *Wildlife Society Bulletin* 33:422–434.

Kaisanlahti-Jokimäki, M.-L., J. Jokimäki, E. Huhta, and P. Siikamäki. 2012. Impacts of seasonal small-scale urbanization on nest predation and bird assemblages at tourist destinations. *Studies in Avian Biology* 45:93–109.

Kaplan, R. 2011. Intrinsic and aesthetic values of urban nature: a psychological perspective. Pages 385–393 in *The Routledge Handbook of Urban Ecology*, edited by I. Douglas, D. Goode, M. C. Houck, and R. Wang. London: Routledge.

Kath, J., M. Maron, and P. K. Dunn. 2009. Interspecific competition and small bird diversity in an urbanizing landscape. *Landscape and Urban Planning* 92:72–79.

Kaye, J. P., P. P. M. Groffman, N. B. Grimm, L. A. Baker, and R. V. Pouyat. 2006. A distinct urban biogeochemistry? *Trends in Ecology and Evolution* 21:193–199.

Keil, A. 2005. Use and perception of post-industrial urban landscapes in the Ruhr. Pages 117–131 in *Wild Urban Woodlands: New Perspectives for Urban Forestry*, edited by I. Kowarik and S. Körner. Berlin: Springer.

Kellert, S. R. 1999. Ecological challenge, human values of nature, and sustainability in the built environment. Pages 39–53 in *Reshaping the Built Environment*, edited by C. J. Kibert. Washington, DC: Island Press.

——. 2012. *Birthright: People and Nature in the Modern World*. New Haven, CT: Yale University Press.

Kellert, S. R., and E. O. Wilson, eds. 1995. *The Biophilia Hypothesis*. Washington, DC: Island Press.

Kendeigh, S. C. 1944. Measurement of bird populations. *Ecological Monographs* 41:68–106.

Kertson, B. N., R. D. Spencer, and C. E. Grue. 2013. Demographic influences on cougar residential use and interactions with people in western Washington. *Journal of Mammalogy* 94:269–281.

Kertson, B. N., R. D. Spencer, J. M. Marzluff, J. Hepinstall Cymerman, and C. E. Grue. 2011. Cougar space use and movements in the wildland-urban landscape of western Washington. *Ecological Applications* 21:2866–2881.

Kinnison, M. T., and A. P. Hendry. 2001. The pace of modern life II: from rates of contemporary microevolution to pattern and process. *Genetica* 112–113:145–164.

Kitchen, L. 2013. Are trees always "good"? Urban political ecology and environmental justice in valleys of south Wales. *International Journal of Urban and Regional Research* 37:1968–1983.

Klem, D., Jr. 2008. Avian mortality at windows: the second largest human source of bird mortality on Earth. *Proceedings of the Fourth International Partners in Flight Conference*, 244–251.

——. 2009. Preventing bird-window collisions. *Wilson Journal of Ornithology* 121: 314–321.

Knight, C. R., and J. P. Swaddle. 2007. Associations of anthropogenic activity and disturbance with fitness metrics of eastern bluebirds (*Sialia sialis*). *Biological Conservation* 138:189–197.

Knight, R. L. 1990. Ecological principles applicable to the management of urban ecosystems. Pages 24–34 in *Perspectives in Urban Ecology: Proceedings of the Symposium,* edited by E. A. Webb and S. Q. Foster. Denver: Denver Museum of Natural History and Thorne Ecological Institute.

Koenig, W. D. 2003. European starlings and their effect on native cavity nesting birds. *Conservation Biology* 17:1134–1140.

Kowarik, I. 1995. On the role of alien species in urban flora and vegetation. Pages 85–103 in *Plant Invasions: General Aspects and Special Problems*, edited by P. Pysek, K. Prach, M. Rejmanek, and M. Wade. Amsterdam: SPB Academic Publishing.

Kowarik, I., and S. Körner, eds. 2005. *Wild Urban Woodlands: New Perspectives for Urban Forestry*. Berlin: Springer.

Krasny, M. E., and K. G. Tidball. 2012. Civic ecology: a pathway for Earth stewardship in cities. *Frontiers in Ecology and the Environment* 10:267–273.

Kübler, S., S. Kupko, and U. Zeller. 2005. The kestrel (*Falco tinnunculus* L.) in Berlin: investigation of breeding biology and feeding ecology. *Journal of Ornithology* 146:271–278.

Kühnel, K.-D. 2008. Railway tracks as habitats for the sand lizard, *Lacerta agilis*, in urban Berlin, Germany. Pages 171–174 in *Urban Herpetology*, edited by J. C. Mitchell, R. E. Jung Brown, and B. Bartholomew. Salt Lake City: Society for the Study of Amphibians and Reptiles.

Kunstler, J. H. 1993. *The Geography of Nowhere*. New York: Simon and Schuster.

Lacheretz, A., D. Moreau, and H. Cathelain. 2002. Causes of death and life expectancy in carnivorous pets (I). *Revue de Médecine Vétérinaire* 12:819–822.

Lampe, U., T. Schmoll, A. Franzke, and K. Reinhold. 2012. Staying tuned: grasshoppers from noisy roadside habitats produce courtship signals with elevated frequency components. *Functional Ecology* 26:1348–1354.

Lande, R. 1976. Natural selection and random genetic drift in phenotypic evolution. *Evolution* 30:314–334.

Lande, R., S. Engen, and B.-E. Saether. 2003. *Stochastic Population Dynamics in Ecology and Conservation*. Oxford: Oxford University Press.

LeClerc, J. E., J. P. K. Che, J. P. Swaddle, and D. A. Cristol. 2005. Reproductive success and developmental stability of eastern bluebirds on golf courses: evidence that golf courses can be productive. *Wildlife Society Bulletin* 33:483–493.

LeClerc, J. E., and D. A. Cristol. 2005. Are golf courses providing habitat for birds of conservation concern in Virginia? *Wildlife Society Bulletin* 33:463–470.

Leopold, A. 1949. *A Sand County Almanac*. Oxford: Oxford University Press.

Lepczyk, C. A., and P. S. Warren, eds. 2012. *Urban Bird Ecology and Conservation*. Berkeley: University of California Press.

Lepczyk, C. A., P. S. Warren, L. Machabée, A. P. Kinzig, and A. G. Mertig. 2012. Who feeds the birds? *Studies in Avian Biology* 45:267–284.

Lerman, S. B., P. S. Warren, H. Gan, and E. Shochat. 2012. Linking foraging decisions to residential yard bird composition. *PLoS One* 7(8): e43497.

Leu, M., S. E. Hanser, and S. T. Knick. 2008. The human footprint in the west: a large-scale analysis of anthropogenic impacts. *Ecological Applications* 18:1119–1139.

Little, R. M., and J. L. Sutton. 2013. Perceptions towards Egyptian Geese at the Steenberg golf estate, Cape Town, South Africa. *Ostrich* 84:85–87.

Livezey, K. B. 2009. Range expansion of barred owls, part I: chronology and distribution. *American Midland Naturalist* 161:49–56.

Longcore, T., C. Rich, P. Mineau, B. MacDonald, D. G. Bert, L. M. Sullivan, E. Mutrie, S. A. Gauthreaux Jr., M. L. Avery, R. L. Crawford, A. M. Manville II, E. R. Travis, and D. Drake. 2012. An estimate of avian mortality at communication towers in the United States and Canada. *PLoS One* 7:e34025.

——. 2013. Avian mortality at communication towers in the United States and Canada: which species, how many, and where? *Biological Conservation* 158:410–419.

Longrich, N. R., T. Tokaryk, and D. J. Field. 2011. Mass extinction of birds at the Cretaceous-Paleogene (K-Pg) boundary. *Proceedings of the National Academy of Sciences (USA)* 108:15253 15257.

Loreau, M. 2000. Biodiversity and ecosystem functioning: recent theoretical advances. *Oikos* 91:3–17.

Loss, S. R., and R. B. Blair. 2011. Reduced density and nest survival of ground-nesting songbirds relative to earthworm invasions in northern hardwood forests. *Conservation Biology* 25:983–992.

Loss, S. R., T. Will, S. S. Loss, and P. P. Marra. 2014. Bird-building collisions in the United States: estimates of annual mortality and species vulnerability. *Condor* 116:8–23.

Loss, S. R., T. Will, and P. P. Marra. 2013. The impact of free-ranging domestic cats on wildlife of the United States. *Nature Communications* 4:1396. DOI: 10.1038/ncomms2380.

Louv, R. 2005. *Last Child in the Woods: Saving Our Children from Nature-Deficit Disorder.* Chapel Hill, NC: Algonquin Books.

Lowry, H., A. Lill, and B. B. M. Wong. 2012. How noisy does a noisy miner have to be? Amplitude adjustments of alarm calls in an avian urban "adapter." *PLoS One* 7:e29960.

Luther, D. A., and E. P. Derryberry. 2012. Birdsongs keep pace with city life: changes in song over time in an urban songbird affects communication. *Animal Behaviour* 83:1059–1066.

Luttik, J. 2000. The value of trees, water and open space as reflected by house prices in The Netherlands. *Landscape and Urban Planning* 48:422–434.

MacGregor-Fors, I. 2008. Relation between habitat attributes and bird richness in a western Mexico suburb. *Landscape and Urban Planning* 84:92–98.

MacGregor-Fors, I., L. Morales-Perez, and J. E. Schondube. 2010. Migrating to the city: responses of neotropical migrant bird communities to urbanization. *Condor* 112: 711–717.

———. 2012. From forests to cities: effects of urbanization on tropical birds. *Studies in Avian Biology* 45:33–49.

Magle, S. B., V. M. Vernon, and K. R. Crooks. 2012. Urban wildlife research: past, present, and future. *Biological Conservation* 155:23–32.

Manville, A. M., II. 2008. Towers, turbines, power lines, and buildings—steps being taken by the U.S. Fish and Wildlife Service to avoid or minimize take of migratory birds at these structures. *Proceedings of the Fourth International Partners in Flight Conference*, 262–272.

Marra, P. P., and R. Reitsma. 2012. Neighborhood nestwatch: mentoring citizens in the urban matrix. Pages 43–50 in *Citizen Science: Public Participation in Environmental Research*, edited by J. L. Dickinson and R. Bonney. Ithaca, NY: Cornell University Press.

Marsh, G. P. 1864 (1985). *Man and Nature; or, Physical Geography as Modified by Human Action*. Cambridge, MA: Belknap Press of Harvard University.

Martin, K., K. E. H. Aitken, and K. L. Wiebe. 2004. Nest sites and nest webs for cavity-nesting communities in interior British Columbia, Canada: nest characteristics and niche partitioning. *Condor* 106:5–19.

Martin, K., and J. M. Eadie. 1999. Nest webs: a community-wide approach to the management and conservation of cavity-nesting forest birds. *Forest Ecology and Management* 115:243–257.

Marzluff, J. M. 2001. Worldwide urbanization and its affects on birds. Pages 19–47 in *Avian Ecology and Conservation in an Urbanizing World*, edited by J. M. Marzluff, R. Bowman, and R. Donnelly. Norwell, MA: Kluwer.

———. 2002. Fringe conservation: a call to action. *Conservation Biology* 16:1175–1176.

———. 2005. Island biogeography for an urbanizing world: how extinction and colonization may determine biological diversity in human-dominated landscapes. *Urban Ecosystems* 8:155–175.

———. 2012. Urban evolutionary ecology. *Studies in Avian Biology* 45:287–308.

Marzluff, J. M., and T. Angell. 2005a. *In the Company of Crows and Ravens.* New Haven, CT: Yale University Press.

———. 2005b. Cultural coevolution: how the human bond with crows and ravens extends theory and raises new questions. *Journal of Ecological Anthropology* 9:67–73.

Marzluff, J. M., and R. P. Balda. 1992. *The Pinyon Jay: Behavioral Ecology of a Colonial and Cooperative Corvid.* London: T&AD Poyser.

Marzluff, J. M., R. Bowman, and R. Donnelly, eds. 2001. *Avian Ecology and Conservation in an Urbanizing World.* Boston: Kluwer Academic Publishers.

Marzluff, J. M., and G. A. Bradley. 2003. Ecological restoration in the settled-wildland interface. Pages 353–370 in *Ecological Restoration of Southwestern Ponderosa Pine Forests,* edited by P. Friederici. Washington, DC: Island Press.

Marzluff, J. M., and K. Ewing. 2001. Restoration of fragmented landscapes for the conservation of birds: a general framework and specific recommendations for urbanizing landscapes. *Restoration Ecology* 9:280–292.

Marzluff, J. M., and N. Hamel. 2001. Land use issues. Pages 659–673 in *Encyclopedia of Biodiversity,* edited by S. A. Levin. San Diego: Academic Press.

Marzluff, J. M., and M. L. Miller. 2014. Crows and crow feeders: observations on interspecific semiotics. Pages 191–211 in *Biocommunication of Animals,* edited by G. Witzany. Dordrecht: Springer.

Marzluff, J. M., and M. Restani. 1999. The effects of forest fragmentation on avian nest predation. Pages 155–169 in *Forest Fragmentation: Wildlife and Management Implications,* edited by J. A. Rochelle, L. A. Lehmann, and J. Wisniewski. Leiden: Brill.

Marzluff, J. M., and A. K. Rodewald. 2008. Conserving biodiversity in urbanizing areas: nontraditional views from a bird's perspective. *Cities and the Environment* 1(2):6. Available at http://escholarship.bc.edu/cate/vol1/iss2/6.

Marzluff, J. M., E. Shulenberger, W. Endlicher, M. Alberti, G. Bradley, C. Ryan, U. Simon, and C. ZumBrunnen, eds. 2008. *Urban Ecology: An International Perspective on the Interaction between Humans and Nature.* New York: Springer.

Marzluff, J. M., J. C. Withey, K. A. Whittaker, M. D. Oleyar, T. M. Unfried, S. Rullman, and J. DeLap. 2007. Consequences of habitat utilization by nest predators and breeding songbirds across multiple scales in an urbanizing landscape. *Condor* 109:516–534.

Mason, R. D. 2008. Impact of swimming pools on amphibians. Pages 271–273 in *Urban Herpetology,* edited by J. C. Mitchell, R. E. Jung Brown, and B. Bartholomew. Salt Lake City: Society for the Study of Amphibians and Reptiles.

Matsuoka, R., and W. Sullivan. 2011. Urban nature: human psychological and community health. Page 408–423 in *The Routledge Handbook of Urban Ecology*, edited by I. Douglas, D. Goode, M. C. Houck, and R. Wang. London: Routledge.

Matteson, K. C., and G. A. Langellotto. 2010. Determinants of inner city butterfly and bee species richness. *Urban Ecosystems* 13:333–347.

McCaffrey, R. E., W. R. Turner, and A. J. Borens. 2012. A new approach to urban bird monitoring: the Tucson Bird Count. *Studies in Avian Biology* 45:139–153.

McDonnell, M. J., A. K. Hahs, and J. H. Breuste, eds. 2009. *Ecology of Cities and Towns: A Comparative Approach*. Cambridge: Cambridge University Press.

McDonnell, M. J., and S. T. A. Pickett, eds. 1993. *Humans as Components of Ecosystems: The Ecology of Subtle Human Effects and Populated Areas*. New York: Springer.

McDonnell, M. J., S. T. A. Pickett, P. Groffman, P. Bohlen, R. V. Pouyat, W. C. Zipperer, R. W. Parmelee, M. M. Carreiro, and K. Medley. 1997. Ecosystem processes along an urban-to-rural gradient. *Urban Ecosystems* 1:21–36.

McFadzen, M. E., and J. M. Marzluff. 1996. Mortality of Prairie Falcons during the post-fledging dependence period. *Condor* 98:791–800.

McIntyre, N. E., and M. E. Hostetler. 2001. Effects of urban land use on pollinator (Hymenopetera: Apoidea) communities in a desert metropolis. *Basic and Applied Ecology* 2:209–218.

McIntyre, N. E., and J. J. Rango. 2009. Arthropods in urban ecosystems: community patterns as functions of anthropogenic land use. Pages 233–242 in *Ecology of Cities and Towns: A Comparative Approach*, edited by M. J. McDonnell, A. K. Hahs, and J. H. Breuste. Cambridge: Cambridge University Press.

McKinney, M. L. 2002. Urbanization, biodiversity, and conservation. *BioScience* 52: 883–890.

———. 2006. Urbanization as a major cause of biotic homogenization. *Biological Conservation* 127:247–260.

McKinney, M. L., and J. L. Lockwood. 1999. Biotic homogenization: a few winners replacing many losers in the next mass extinction. *Trends in Ecology and Evolution* 14:450–453.

Meffert, P. J. M. Marzluff, and F. Dziock. 2012. Unintentional habitats: value of a city for the wheatear (*Oenanthe oenanthe*). *Landscape and Urban Planning* 108:49–56.

Melles, S. 2000. *Effects of Landscape vs. Local Habitat Features on Bird Communities: A Study of an Urban Gradient in Greater Vancouver*. Master's thesis. Vancouver, University of British Columbia.

References

Melles, S., S. Glenn, and K. Martin. 2003. Urban bird diversity and landscape complexity: species-environment associations along a multiscale habitat gradient. *Conservation Ecology* 7(1):5. Available at http://www.ecologyandsociety.org/vol7/iss1/art5/main.html.

Merola-Zwartjes, M., and J. P. DeLong. 2005. Avian species assemblages on New Mexico golf courses: surrogate riparian habitat for birds? *Wildlife Society Bulletin* 33:435–447.

Meurk, C. D., N. Zvyagna, R. O. Gardner, G. Forrester, M. Wilcox, G. Hall, H. North, S. Belliss, K. Whaley, B. Sykes, J. Cooper, and K. O'Halloran. 2009. Environmental, social and spatial determinants of urban arboreal character in Auckland, New Zealand. Pages 287–307 in *Ecology of Cities and Towns: A Comparative Approach*, edited by M. J. McDonnell, A. K. Hahs, and J. H. Breuste. Cambridge: Cambridge University Press.

Meyers, L. A., S. Sullivan, and P. Mazeika. 2013. Bright lights, big city: influences of ecological light pollution on reciprocal stream-riparian invertebrate fluxes. *Ecological Applications* 23:1322–1330.

Mifsud, D. A., and R. Mifsud. 2008. Golf courses as refugia for herptofauna in an urban river floodplain. Pages 303–310 in *Urban Herpetology*, edited by J. C. Mitchell, R. E. Jung Brown, and B. Bartholomew. Salt Lake City: Society for the Study of Amphibians and Reptiles.

Milder, J. C. 2007. A framework for understanding conservation development and its ecological implications. *BioScience* 57:757–768.

Milesi, C., S. W. Running, C. D. Elvidge, J. B. Dietz, B. T. Tuttle, and R. R. Nemani. 2005. Mapping and modeling the biogeochemical cycling of turf grasses in the United States. *Environmental Management* 36:426–438.

Miller, J. R. 2005. Biodiversity conservation and the extinction of experience. *Trends in Ecology and Evolution* 20:430–434.

———. 2006. Restoration, reconciliation, and reconnecting with nature nearby. *Biological Conservation* 127:356–361.

Miller, J. R., and R. J. Hobbs. 2002. Conservation where people live and work. *Conservation Biology* 16:330–337.

Milus, S. 2013. Collision course. *Science News* 184:20–25.

Mineau, P., and C. Palmer. 2013. *The Impact of the Nation's Most Widely Used Insecticides on Birds*. Washington, DC: American Bird Conservancy.

Mitchell, J. C., and R. E. Jung Brown. 2008. Urban herpetology: global overview, synthesis, and future directions. Pages 1–30 in *Urban Herpetology*, edited by J. C.

Mitchell, R. E. Jung Brown, and B. Bartholomew. Salt Lake City: Society for the Study of Amphibians and Reptiles.

Mitchell, J. C., R. E. Jung Brown, and B. Bartholomew, eds. 2008. *Urban Herpetology*. Salt Lake City: Society for the Study of Amphibians and Reptiles.

Molino, J.-F., and D. Sabatier. 2001. Tree diversity in tropical rain forests: a validation of the intermediate disturbance hypothesis. *Science* 294:1702–1704.

Mooney, H. A., and R. J. Hobbs, eds. 2000. *Invasive Species in a Changing World*. Washington, DC: Island Press.

Moore, W. S. 1995. Northern flicker, *Colaptes auratus*. *Birds of North America* 166:1–27.

Mora, C., and P. F. Sale. 2011. Ongoing global biodiversity loss and the need to move beyond protected areas: a review of the technical and practical shortcomings of protected areas on land and sea. *Marine Ecology Progress Series* 434:251–266.

Murton, R. K., R. J. P. Thearle, and C. F. B. Coombs. 1974. Ecological studies of the feral pigeon *Columba livia* var. III. Reproduction and plumage polymorphism. *Journal of Applied Ecology* 11:841–854.

Murton, R. K., N. J. Westwood, and R. J. P. Thearle. 1973. Polymorphism and the evolution of a continuous breeding season in the pigeon, *Columba livia*. *Journal of Reproduction and Fertility Supplement* 19:563–577.

Naeem, S. 2002. Ecosystem consequences of biodiversity loss: the evolution of a paradigm. *Ecology* 83:1537–1552.

Natuhara, Y., and H. Hashimoto. 2009. Spatial pattern and process in urban animal communities. Pages 197–214 in *Ecology of Cities and Towns: A Comparative Approach*, edited by M. J. McDonnell, A. K. Hahs, and J. H. Breuste. Cambridge: Cambridge University Press.

Nicholson-Lord, D. 2011. Intrinsic and aesthetic values of urban nature: a journalist's view from London. Pages 377–384 in *The Routledge Handbook of Urban Ecology*, edited by I. Douglas, D. Goode, M. C. Houck, and R. Wang. London: Routledge.

Norris, J. L. 2011. *Urban Development in Costa Rica: The Direct and Indirect Impacts on Local and Regional Avian Assemblages*. PhD diss. St. Louis, University of Missouri–St. Louis. Available at https://apps.umsl.edu/webapps/weboffice/ETD/query.cfm?id=r7761.

Oleyar, M. D. 2011. *Urbanization Influences on Songbird Population Dynamics, Community Structure, and Energy Relationships*. PhD diss. Seattle, University of Washington.

Oleyar, M. D., A. I. Greve, J. C. Withey, and A. M. Bjorn. 2008. An integrated approach to evaluating urban forest functionality. *Urban Ecosystems* 11:289–308.

Ostergaard, E. C., K. O. Richter, and S. D. West. 2008. Amphibian use of stormwater ponds in the Puget lowlands of Washington, USA. Pages 259–270 in *Urban Herpetology*, edited by J. C. Mitchell, R. E. Jung Brown, and B. Bartholomew. Salt Lake City: Society for the Study of Amphibians and Reptiles.

Paine, R. T. 1966. Food web complexity and species diversity. *American Naturalist* 100:65–75.

———. 1980. Food webs: linkage, interaction strength and community infrastructure. *Journal of Animal Ecology* 49:667–685.

Paker, Y., Y. Yom-Tov, T. Alon-Mozes, and A. Barnea. 2013. The effect of plant richness and urban garden structure on bird species richness, diversity and community structure. *Landscape and Urban Planning* 122:186–195.

Paloski, R. A. 2008. Relationship between lakeshore development and frog populations of central Wisconsin. Pages 77–83 in *Urban Herpetology*, edited by J. C. Mitchell, R. E. Jung Brown, and B. Bartholomew. Salt Lake City: Society for the Study of Amphibians and Reptiles.

Palumbi, S. R. 2001. *The Evolution Explosion: How Humans Cause Rapid Evolutionary Change*. New York: W. W. Norton.

Parris, K. M., and D. L. Hazell. 2005. Biotic effects of climate change in urban environments: the case of the grey-headed flying-fox (*Pteropus poliocephalus*) in Melbourne, Australia. *Biological Conservation* 124:267–276.

Parris, K. M., M. Velik-Lord, and J. M. A. North. 2009. Frogs call at a higher pitch in traffic noise. *Ecology and Society* 14(1):25. Available at http://www.ecologyandsociety.org/vol14/iss1/art25/main.html.

Partecke, J., E. Gwinner, and S. Bensch. 2006a. Is urbanization of European blackbirds (*Turdus merula*) associated with genetic differentiation? *Journal of Ornithology* 147:549–552.

Partecke, J., I. Schwabl, and E. Gwinner. 2006b. Stress and the city: urbanization and its effects on the stress physiology in European blackbirds. *Ecology* 87:1945–1952.

Partecke, J., T. Van't Hof, and E. Gwinner. 2004. Differences in the timing of reproduction between urban and forest European blackbirds (*Turdus merula*): result of phenotypic flexibility or genetic differences? *Proceedings of the Royal Society of London B* 271:1995–2001.

Paton, P. W. C., C. McDonough, and K. E. Montieth. 2008. Migration ecology of spotted salamanders (*Ambystoma maculatum*) on golf courses in southern New England. Pages 293–301 in *Urban Herpetology*, edited by J. C. Mitchell, R. E. Jung Brown, and B. Bartholomew. Salt Lake City: Society for the Study of Amphibians and Reptiles.

Patricelli, G. L., and J. L. Blickley. 2006. Avian communication in urban noise: causes and consequences of vocal adjustment. *Auk* 123:639–649.

Pennington, D. N., and R. B. Blair. 2011. Habitat selection of breeding riparian birds in an urban environment: untangling the relative importance of biophysical elements and spatial scale. *Diversity and Distributions* 17:506–518.

———. 2012. Using gradient analysis to uncover pattern and process in urban bird communities. *Studies in Avian Biology* 45:9–31.

Perry, G., B. W. Buchanan, R. N. Fisher, M. Salmon, and S. E. Wise. 2008. Effects of artificial night lighting on amphibians and reptiles in urban environments. Pages 239–256 in *Urban Herpetology*, edited by J. C. Mitchell, R. E. Jung Brown, and B. Bartholomew. Salt Lake City: Society for the Study of Amphibians and Reptiles.

Pickett, S. T. A., M. L. Cadenasso, J. M. Grove, C. H. Nilon, R. V. Pouyat, W. C. Zipperer, and R. Costanza. 2001. Urban ecological systems: linking terrestrial ecological, physical, and socioeconomic components of metropolitan areas. *Annual Review of Ecology and Systematics* 21:127–157.

Pickett, S. T. A., M. L. Cadenasso, M. J. McDonnell, and W. R. Burch Jr. 2009. Frameworks for urban ecosystem studies: gradients, patch dynamics and the human ecosystem in the New York metropolitan area and Baltimore, USA. Pages 25–50 in *Ecology of Cities and Towns: A Comparative Approach*, edited by M. J. McDonnell, A. K. Hahs, and J. H. Breuste. Cambridge: Cambridge University Press.

Pimm, S. L., M. P. Moulton, and L. J. Justice. 1994. Bird extinctions in the central Pacific. *Philosophical Transactions of the Royal Society of London B* 344:27–33.

Pimm, S. L., G. J. Russell, J. L. Gitteleman, and T. M. Brooks. 1995. The future of biodiversity. *Science* 269:347–350.

Plöger, J. 2012. Learning from abroad: lessons from European shrinking cities. Pages 295–321 in *Rebuilding America's Legacy Cities: New Directions for the Industrial Heartland*, edited by A. Mallach. New York: American Assembly, Columbia University.

Polis, G. A., and D. R. Strong. 1996. Food web complexity and community dynamics. *American Naturalist* 147:813–846.

Poot, H., B. J. Ens, H. de Vries, M. A. H. Donners, M. Wernand, and J. M. Marquenie. 2008. Green light for nocturnally migrating birds. *Ecology and Society* 13(2):47. Available at http://www.ecologyandsociety.org/vol13/iss2/art47/main.html.

Pope, A. 1996. *Ladders*. Houston: Rice University School of Architecture.

Porter, E. E., J. Bulluck, and R. B. Blair. 2005. Multiple spatial-scale assessment of the conservation value of golf courses for breeding birds in southwestern Ohio. *Wildlife Society Bulletin* 33:494–506.

Potvin, D. A., K. M. Parris, and R. A. Mulder. 2013. Limited genetic differentiation between acoustically divergent populations of urban and rural silvereyes (*Zosterops lateralis*). *Evolutionary Ecology* 27:381–391.

Pouliot, Y. 2008. Les collisions d'oiseaux à l'édifice Marly à Sainte-Foy, Québec, de 1978 à 2007. *Canadian Field Naturalist* 122:153–157.

Price, T. D., A. Qvarnström, and D. E. Irwin. 2003. The role of phenotypic plasticity in driving genetic evolution. *Proceedings of the Royal Society of London B* 270:1433–1440.

Proppe, D. S., C. B. Sturdy, and C. C. St. Claire. 2013. Anthropogenic noise decreases urban songbird diversity and may contribute to homogenization. *Global Change Biology* 19:1075–1084.

Pyle, P. 1997. *Identification Guide to North American Birds, Part 1.* Bolinas, CA: Slate Creek Press.

Qian, Y., and R. F. Follett. 2002. Assessing soil carbon sequestration in turfgrass systems using long-term soil testing data. *Agronomy Journal* 94:930–934.

Quinn, T. 1997. Coyote (*Canis latrans*) food habits in three urban habitat types of western Washington. *Northwest Science* 71:1–5.

Rainwater, T. R., V. A. Leopold, M. J. Hooper, and R. J. Kendall. 1995. Avian exposure to organophosphorus and carbamate pesticides on a coastal South Carolina golf course. *Environmental Toxicology and Chemistry* 14:2155–2161.

Ralph, C. J., G. R. Geupel, P. Pyle, T. E. Martin, and D. F. DeSante. 1993. Handbook of field methods for monitoring landbirds. General Technical Report PSW-GTR-144. Albany, CA: U.S. Department of Agriculture, Forest Service, Pacific Southwest Research Station.

Rebele, F. 1994. Urban ecology and special features of urban ecosystems. *Global Ecology and Biogeography Letters* 4:173–187.

Recher, H. F., and D. L. Serventy. 1991. Long term changes in the relative abundances of birds in Kings Park, Perth, Western Australia. *Conservation Biology* 5:90–102.

Redford, K. H., and B. D. Richter. 1999. Conservation of biodiversity in a world of use. *Conservation Biology* 13:1246–1256.

Reznick, D. N., and C. K. Ghalambor. 2001. The population ecology of contemporary adaptations: what empirical studies reveal about the conditions that promote adaptive evolution. *Genetica* 112–113:183–198.

Rich, C., and T. Longcore, eds. 2006. *Ecological Consequences of Artificial Night Lighting*. Washington, DC: Island Press.

Richards, J. F. 1990. Land transformations. Pages 163–178 in *The Earth as Transformed by Human Action*, edited by B. L. Turner II, W. C. Clark, R. W. Kates, J. F. Richards, J. T. Mathews, and W. B. Meyer. Cambridge: Cambridge University Press.

Ricketts, T., and M. Imhoff. 2003. Biodiversity, urban areas, and agriculture: locating priority ecoregions for conservation. *Conservation Ecology* 8(2):1. Available at http://www.ecologyandsociety.org/vol8/iss2/art1/main.html.

Riem, J. G., R. B. Blair, D. N. Pennington, and N. G. Solomon. 2012. Estimating mammalian species diversity across an urban gradient. *American Midland Naturalist* 168:315–332.

Riley, S. P. D., J. P. Pollinger, R. M. Sauvajot, E. C. York, D. A. Kamradt, C. Bromley, T. K. Fuller, and R. K. Wayne. 2006. A southern California freeway is a physical and social barrier to gene flow in carnivores. *Molecular Ecology* 15:1733–1741.

Riley, S. P. D., R. M. Sauvajot, T. K. Fuller, E. C. York, D. A. Kamradt, C. Bromley, and R. K. Wayne. 2003. Effects of urbanization and habitat fragmentation on bobcats and coyotes in southern California. *Conservation Biology* 17:566–576.

Robb, G. N., R. A. McDonald, D. E. Chamberlain, J. Reynolds, T. J. E. Harrison, and S. Bearhop. 2008b. Winter feeding of birds increases productivity in the subsequent breeding season. *Biology Letters* 4:220–223.

Robb, G. N., R. A. McDonald, D. E. Chamberlain, and S. Bearhop. 2008a. Food for thought: supplementary feeding as a driver of ecological change in avian populations. *Frontiers in Ecology the Environment* 6:476–484.

Robbins, P. 2007. *Lawn People: How Grasses, Weeds, and Chemicals Make Us Who We Are*. Philadelphia: Temple University Press.

Robertson, P. B., and A. F. Schnapf. 1987. Pyramiding behavior in the Inca dove—adaptive aspects of day-night differences. *Condor* 89:185–187.

Robinson, L., J. P. Newell, and J. M. Marzluff. 2005. Twenty-five years of sprawl in the Seattle region: growth management responses and implications for conservation. *Landscape and Urban Planning* 71:51–72.

Rodewald, A. D. 2012. Evaluating factors that influence avian community response to urbanization. *Studies in Avian Biology* 45:71–92.

Rodewald, A. D., and D. P. Shustack. 2008. Urban flight: understanding individual and population-level responses of Nearctic-Neotropical migratory birds to urbanization. *Journal of Animal Ecology* 77:83–91.

References

Rodewald, A. D., D. P. Shustack, and L. E. Hitchcock. 2010. Exotic shrubs as ephemeral ecological traps for nesting birds. *Biological Invasions* 12:33–39.

Rodewald, P. G., M. J. Santiago, and A. D. Rodewald. 2005. Habitat use of breeding red-headed woodpeckers on golf courses in Ohio. *Wildlife Society Bulletin* 33:448–453.

Rolshausen, G., G. Segelbacher, K. A. Hobson, and H. M. Schaefer. 2009. Contemporary evolution of reproductive isolation and phenotypic divergence in sympatry along a migratory divide. *Current Biology* 19:2097–2101.

Root, T., and S. H. Schneider. 2006. Conservation and climate change: the challenges ahead. *Conservation Biology* 20:706–708.

Rosenberg, K. V., S. B. Terrill, and G. H. Rosenberg. 1987. Value of suburban habitats to desert riparian birds. *Wilson Bulletin* 99:642–654.

Rosenfield, R. N., J. Bielefeldt, J. L. Affeldt, and D. J. Beckmann. 1996. Urban nesting biology of Cooper's hawks in Wisconsin. Pages 41–44 in *Raptors in Human Landscapes: Adaptations to Built and Cultivated Environments*, edited by D. Bird, D. Varland, and J. Negro. London: Academic Press.

Rotenberg, J. A., L. M. Barnhill, J. M. Meyers, and D. Demarest. 2012. Painted bunting conservation: traditional monitoring meets citizen science. *Studies in Avian Biology* 45:125–137.

Rullman, S. D., Jr. 2012. *Urban Raptors and Patterns of Second Home Development: Interdisciplinary Studies in Urban Ecology.* PhD diss. Seattle, University of Washington.

Rullman, S. D., Jr. and J. M. Marzluff. 2014. Raptor presence along an urban–wildland gradient: influence of prey abundance and land cover. *Journal of Raptor Research* (in press).

Ryder, T. B., R. Reitsma, B. Evans, and P. P. Marra. 2010. Quantifying avian nest survival along an urbanization gradient using citizen and scientist generated data. *Ecological Applications* 20:419–426.

Sattler, T., D. Borcard, R. Arlettaz, F. Bontadina, P. Legendre, M. K. Obrist, and M. Moretti. 2010. Spider, bee and bird communities in cities are shaped by environmental control and high stochasticity. *Ecology* 91:3343–3353.

Sauer, J. R., and S. Droege. 1990. Recent population trends of the eastern bluebird. *Wilson Bulletin* 102:239–252.

Sauvajot, R. M., M. Buechner, D. A. Kamradt, and C. M. Schonewald. 1998. Patterns of human disturbance and response by small mammals and birds in chaparral near urban development. *Urban Ecosystems* 2:279–297.

Schepps, J., S. Lohr, and T. E. Martin. 1999. Does tree hardness influence nest-tree selection by primary cavity nesters? *Auk* 116:658–665.

Schimel, D. S., G. P. Asner, and P. Moorcroft. 2013. Observing changing ecological diversity in the Anthropocene. *Frontiers in Ecology and the Environment* 11:129–137.

Schindler, D. E., S. I. Geib, and M. R. Williams. 2000. Patterns of fish growth along a residential development gradient in north temperate lakes. *Ecosystems* 3:229–237.

Schmidt, B. R., and S. Zumbach. 2008. Amphibian road mortality and how to prevent it: a review. Pages 157–167 in *Urban Herpetology*, edited by J. C. Mitchell, R. E. Jung Brown, and B. Bartholomew. Salt Lake City: Society for the Study of Amphibians and Reptiles.

Schoech, S. J., and R. Bowman. 2001. Variation in the timing of breeding between suburban and wildland Florida Scrub-jays: do physiologic measures reflect different environments? Pages 289–306 in *Avian Ecology and Conservation in an Urbanizing World*, edited by J. M. Marzluff, R. Bowman, and R. Donnelly. Boston: Kluwer Academic Publishers.

——.2003. Does differential access to protein influence differences in timing of breeding of Florida Scrub-jays (*Aphelocoma coerulescens*) in suburban and wildland habitats? *Auk* 120:1114–1127.

Scott, D. E., B. S. Metts, and J. W. Gibbons. 2008. Enhancing amphibian biodiversity on golf courses with seasonal wetlands. Pages 285–292 in *Urban Herpetology*, edited by J. C. Mitchell, R. E. Jung Brown, and B. Bartholomew. Salt Lake City: Society for the Study of Amphibians and Reptiles.

Sedláček, O., R. Fuchs, and A. Exnerová. 2004. Redstart *Phoenicurus phoenicurus* and black redstart *P. ochruros* in a mosaic urban environment: neighbors or rivals? *Journal of Avian Biology* 35:336–343.

Seideman, D. 2013. Tornado watch. *Audubon*, July–August:54–56.

Seiler, A. 2003. *The Toll of the Automobile: Wildlife and Roads in Sweden.* PhD diss. Uppsala, Swedish University of Agricultural Sciences.

Selander, R. K., and R. F. Johnston. 1967. Evolution in the house sparrow I. Intrapopulation variation in North America. *Condor* 69:217–258.

Sewell, S. R., and C. P. Catterwall. 1998. Bushland modification and styles of urban development: their effects on birds in south east Queensland. *Wildlife Research* 25:41–63.

Shafer, C. L. 1997. Terrestrial nature reserve design at the urban/rural interface. Pages 345–378 in *Conservation in Highly Fragmented Landscapes*, edited by M. W. Schwartz. New York: Chapman and Hall.

Shaw, L. M., D. Chamberlain, and M. Evans. 2008. The house sparrow *Passer domesticus* in urban areas: reviewing a possible link between post-decline distribution and human socioeconomic status. *Journal of Ornithology* 149:293–299.

Shochat, E., S. B. Lerman, J. M. Anderies, P. S. Warren, S. H. Faeth, and C. H. Nilon. 2010. Invasion, competition, and biodiversity loss in urban ecosystems. *BioScience* 60:199–208.

Shochat, E., P. S. Warren, S. H. Faeth, N. E. McIntyre, and D. Hope. 2006. From pattern to emerging processes in mechanistic urban ecology. *Trends in Ecology and Evolution* 21:186–191.

Slabbekoorn, H., and E. A. P. Ripmeester. 2008. Birdsong and anthropogenic noise: implications and applications for conservation. *Molecular Ecology* 17:72–83.

Slabbekoorn, H., P. Yeh, and K. Hunt. 2007. Sound transmission and song divergence: a comparison of urban and forest acoustics. *Condor* 109:67–78.

Slagsvold, T. 1980. Habitat selection in birds: on the presence of other bird species with special regard to *Turdus pilaris*. *Journal of Animal Ecology* 49:523–536.

Sloane, S. A. 1996. Incidence and origins of supernumeraries at bushtit (*Psaltriparus minimus*) nests. *Auk* 113:757–770.

Smith, J. J., M. Goode, and M. Amarello. 2008. Changes in structure and composition of Sonoran desert reptile communities associated with golf courses. Pages 311–320 in *Urban Herpetology*, edited by J. C. Mitchell, R. E. Jung Brown, and B. Bartholomew. Salt Lake City: Society for the Study of Amphibians and Reptiles.

Smith, M. D., C. J. Conway, and L. A. Ellis. 2005. Burrowing owl nesting productivity: a comparison between artificial and natural burrows on and off golf courses. *Wildlife Society Bulletin* 33:454–462.

Smith, S. B., J. E. McKay, J. K. Richardson, and M. T. Murphy. 2012. Edges, trails, and reproductive performance of spotted towhees in urban greenspaces. *Studies in Avian Biology* 45:167–181.

Snep, R. P. H. 2009. *Biodiversity Conservation at Business Sites*. PhD diss. Alterra Scientific Contributions 28. Wageningen, The Netherlands: Alterra.

Sol, D., O. Lapiedra, and C. González-Lagos. 2013. Behavioural adjustments for a life in the city. *Animal Behaviour* 85:1101–1112.

Sorace, A., and M. Visentin. 2007. Avian diversity on golf courses and surrounding landscapes in Italy. *Landscape and Urban Planning* 81:81–90.

Soulé, M. E. 1991. Land use planning and wildlife maintenance. *Journal of the American Planning Association* 57:313–323.

Soulé, M. E., D. T. Bolger, A. C. Alberts, J. Wright, M. Sorice, and S. Hill. 1988. Reconstructed dynamics of rapid extinctions of chaparral-requiring birds in urban habitat islands. *Conservation Biology* 2:75–92.

Southerland, M. T., and S. A. Stranko. 2008. Fragmentation of riparian amphibian distributions by urban sprawl in Maryland, USA. Pages 423–433 in *Urban Herpetology*, edited by J. C. Mitchell, R. E. Jung Brown, and B. Bartholomew. Salt Lake City: Society for the Study of Amphibians and Reptiles.

Soutullo, A. 2010. Extent of the global network of terrestrial protected areas. *Conservation Biology* 24:362–363.

Stacey, P. B., and M. L. Taper. 1992. Environmental variation and the persistence of small populations. *Ecological Applications* 2:18–29.

Stachowicz, J. J. 2001. Mutualism, facilitation, and the structure of ecological communities. *BioScience* 51:235–246.

Stanback, M. T., and M. L. Seifert. 2005. A comparison of eastern bluebird reproductive parameters in golf and rural habitats. *Wildlife Society Bulletin* 33:471–482.

Steeger, C., and C. L. Hitchcock. 1998. Influence of forest structure and diseases on nest-site selection by red-breasted nuthatches. *Journal of Wildlife Management* 62:1349–1358.

Sterba, J. 2012. *Nature Wars*. New York: Crown.

Stiebens, V. A., S. E. Merino, C. Roder, F. J. Chain, P. L. M. Lee, and C. Eizaguirre. 2013. Living on the edge: how philopatry maintains adaptive potential. *Proceedings of the Royal Society of London B* 280:20130305.

Stockwell, C. A., A. P. Hendry, and M. T. Kinnison. 2003. Contemporary evolution meets conservation biology. *Trends in Ecology and Evolution* 18:94–101.

Stracey, C. M., and S. K. Robinson. 2012. Does nest predation shape urban bird communities? *Studies in Avian Biology* 45:49–70.

Stroud, E. 2012. *Nature Next Door: Cities and Trees in the American Northeast*. Seattle: University of Washington Press.

Tanner, R. A., and A. C. Gange. 2005. Effects of golf courses on local biodiversity. *Landscape and Urban Planning* 71:137–146.

Taratalos, J., R. A. Fuller, K. L. Evans, R. G. Davies, S. E. Newson, J. J. D. Greenwood, and K. J. Gaston. 2007. Bird densities are associated with household densities. *Global Change Biology* 13:1685–1695.

Tarkhnishvili, D. N., and R. K. Gokhelasuili. 1996. A contribution to the ecological genetics of frogs: age structure and frequency of striped specimens in some Caucasian populations of the *Rana macronemis* complex. *Alytes* 14:27–41.

Terborgh, J., L. Lopez, P. Nuñez, M. Rao, G. Shahabuddin, G. Orihuela, M. Riveros, R. Ascanio, G. H. Adler, T. D. Lambert, and L. Balbas. 2001. Ecological meltdown in predator-free forest fragments. *Science* 294:1923–1926.

Terman, M. R. 1996. The bird communities of Prairie Dunes country club and Sand Hills state park. *USGA Green Section Record*, November/December:10–14.

Terrill, S. B., and P. Berthold. 1990. Ecophysiological aspects of rapid population growth in a novel migratory blackcap (*Sylia atricapilla*) population: an experimental approach. *Oecologia* 85:266–270.

Theobald, D. M. 2005. Landscape patterns of exurban growth in the USA from 1980 to 2020. *Ecology and Society* 10(1):32. Available at http://www.ecologyandsociety.org/vol10/iss1/art32.

Thompson, J. N. 1998. Rapid evolution as an ecological process. *Trends in Ecology and Evolution* 13:329–332.

Thompson, R., R. Hanna, J. Noel, and D. Piirto. 1999. Valuation of tree aesthetics on small urban-interface properties. *Journal of Arboriculture* 25:225–234.

Tilt, J. H. 2011. Urban nature and human physical health. Page 394–407 in *The Routledge Handbook of Urban Ecology*, edited by I. Douglas, D. Goode, M. C. Houck, and R. Wang. London: Routledge.

Toms, M. 2003. *The BTO/CJ Garden Birdwatch Book*. Thetford, UK: British Trust for Ornithology.

Trenham, P. C., and D. G. Cook. 2008. Distribution of migrating adults related to the location of remnant grassland around an urban California tiger salamander (*Ambystoma californiense*) breeding pool. Pages 33–40 in *Urban Herpetology*, edited by J. C. Mitchell, R. E. Jung Brown, and B. Bartholomew. Salt Lake City: Society for the Study of Amphibians and Reptiles.

Trut, L. N. 1999. Early canid domestication: the farm-fox experiment: foxes bred for tamability in a 40-year experiment exhibit remarkable transformations that suggest an interplay between behavioral genetics and development. *American Scientist* 87:160–169.

Turner, W. R., T. Nakamura, and M. Dinetti. 2004. Global urbanization and the separation of humans from nature. *BioScience* 54:585–590.

Unfried, T. M. 2009. *Urban Landscape Relationships with Song Sparrow (Melospiza melodia) Population Structure and Connectivity and Human Health and Behavior*. PhD diss. Seattle, University of Washington.

Unfried, T. M., L. Hauser, and J. M. Marzluff. 2013. Effects of urbanization on song sparrow (*Melospiza melodia*) population connectivity. *Conservation Genetics* 14:41–53.

United Nations. 2007. *World Urbanization Prospects: The 2007 Revision.* New York: United Nations.

———. 2012. *World Urbanization Prospects: The 2011 Revision.* New York: United Nations.

U.S. Department of Transportation. 2008. *Wildlife-Vehicle Collision Reduction Study, Report to Congress.* Washington, DC: USDOT, Federal Highway Administration.

U.S. Environmental Protection Agency. 2013. Land use overview. Available at http://www.epa.gov/agriculture/ag101/landuse.html.

U.S. Fish and Wildlife Service. 2005. Draft revised Recovery Plan for the Aga or Mariana Crow. Portland, OR: USFWS. Available at http://endangered.fws.gov/recovery/index.html.

Van der Ree, R. 2009. The ecology of roads in urban and urbanizing landscapes. Page 185–196 in *Ecology of Cities and Towns: A Comparative Approach,* edited by M. J. McDonnell, A. K. Hahs, and J. H. Breuste. Cambridge: Cambridge University Press.

Van Gelder, S. 1999. Environmental ethics. Pages 54–72 in *Reshaping the Built Environment: Ecology, Ethics, and Economics,* edited by C. J. Kibert. Washington, DC: Island Press.

Vargo, T. L., O. D. Boyle, C. A. Lepczyk, W. P. Mueller, and S. E. Vondrachek. 2012. The use of citizen volunteers in urban bird research. *Studies in Avian Biology* 45:113–124.

Vitousek, P. M. 1994. Beyond global warming: ecology and global change. *Ecology* 75:1861–1876.

Walasz, K. 1990. Experimental investigations on the behavioural differences between urban and forest blackbirds. *Acta Zoologica Cracoviensia* 33:235–271.

Walker, J. S., N. B. Grimm, J. M. Briggs, C. Gries, and L. Dugan. 2009. Effects of urbanization on plant species diversity in central Arizona. *Frontiers in Ecology and the Environment* 7:465–470.

Walther, G. R., L. Hughes, P. M. Vitousek, and N. C. Stenseth. 2005. Consensus on climate change. *Trends in Ecology and Evolution* 20:648–649.

Wandeler, P., S. M. Funk, C. R. Largiadèr, S. Gloors, and U. Breitenmoser. 2003. The city-fox phenomenon: genetic consequences of a recent colonization of urban habitat. *Molecular Ecology* 12:647–656.

Warren, P. S., M. Katti, M. Ermann, and A. Brazel. 2006. Urban bioacoustics: it's not just noise. *Animal Behaviour* 71:491–502.

Wells, N. M., and K. S. Lekies. 2012. Children and nature: following the trail to environmental attitudes and behavior. Pages 201–213 in *Citizen Science: Public Participation in Environmental Research,* edited by J. L. Dickinson and R. Bonney. Ithaca, NY: Cornell University Press.

Wheater, C. P. 2011. Walls and paved surfaces. Pages 239–251 in *The Routledge Handbook of Urban Ecology*, edited by I. Douglas, D. Goode, M. C. Houck, and R. Wang. London: Routledge.

Whipple, S. D., and W. W. Hoback. 2012. A comparison of dung beetle (Coleoptera: Scarabaeidae) attraction to native and exotic mammal dung. *Environmental Entomology* 41:238–244.

White, C. L., and M. B. Main. 2005. Waterbird use of created wetlands in golf-course landscapes. *Wildlife Society Bulletin* 33:411–421.

White, G. C., and K. P. Burnham. 1999. Program MARK: survival estimation from populations of marked animals. *Bird Study Supplement* 46:S120–S139.

White, R. L., T. J. Baptiste, A. Dornelly, M. N. Morton, M. J. O'Connell, and R. P. Young. 2012. Population responses of the endangered white-breasted thrasher *Ramphocinclus brachyurus* to a tourist development in Saint Lucia: conservation implications from a spatial modeling approach. *Bird Conservation International* 22:468–485.

Whittaker, K. A. 2007. *Dispersal, Habitat Use, and Survival of Native Forest Songbirds in an Urban Landscape*. PhD diss. Seattle, University of Washington.

Whittaker, K. A., and J. M. Marzluff. 2009. Species-specific survival and relative habitat use in an urban landscape during the postfledging period. *Auk* 126:288–299.

———. 2012. Post-fledging mobility in an urban landscape. *Studies in Avian Biology* 45:183–198.

Wiens, J. A. 1977. On competition and variable environments. *American Scientist* 65:590–597.

Williams, T. 2009. Feline fatales. *Audubon*, September–October. Available at http://archive.audubonmagazine.org/incite/incite0909.html.

Wilson, E. O. 2002. *The Future of Life*. New York: Knopf.

Wilson, S. G., D. A. Plane, P. J. Mackun, T. R. Fischetti, and J. Goworowska. 2012. Patterns of metropolitan and micropolitan population change: 2000 to 2010. Washington, DC: U.S. Census Bureau.

Wiltschko, W., and R. Wiltschko. 2001. Light-dependent magnetoreception in birds: the behavior of European robins, *Erithacus rubecula*, under monochromatic light of various wavelengths and intensities. *Journal of Experimental Biology* 204:3295–3302.

Windmiller, B. S., R. N. Homan, J. V. Regosin, L. A. Willetts, D. L. Wells, and J. M. Reed. 2008. Breeding amphibian population declines following loss of upland forest habitat around vernal pools in Massachusetts, USA. Pages 41–51 in *Urban Herpetology*, edited by J. C. Mitchell, R. E. Jung Brown, and B. Bartholomew. Salt Lake City: Society for the Study of Amphibians and Reptiles.

With, K. A., and T. O. Crist. 1995. Critical thresholds in species responses to landscape structure. *Ecology* 76:2446–2459.

Woodward, I., and R. Arnold. 2012. The changing status of the breeding birds of the Inner London area. *British Birds*. Available at http://www.britishbirds.co.uk/articles/the-changing-status-of-the-breeding-birds-of-the-inner-london-area.

Work, T. M., J. G. Massey, D. S. Lindsay, and J. P. Dubey. 2002. Toxoplasmosis in three species of native and introduced Hawaiian birds. *Journal of Parasitology* 88:1040–1042.

Yan, C. K. 2013. *The Effects of Urban Greenery on Biodiversity*. PhD diss. National University of Singapore.

Yeh, P. J. 2004. Rapid evolution of a sexually selected trait following population establishment in a novel habitat. *Evolution* 58:166–174.

Yeh, P., and T. D. Price. 2004. Adaptive phenotypic plasticity and the successful colonization of a novel environment. *American Naturalist* 164:531–542.

Index

Note: Italic page numbers refer to illustrations and n refers to notes

Index